Glory Days
of
LOGGING

Glory Days of Logging

by

Ralph W. Andrews

77 Lower Valley Road, Atglen, PA 19310

Printed in the United States of America.
ISBN: 0-88740-593-2

Published by Schiffer Publishing, Ltd.
77 Lower Valley Road
Atglen, PA 19310
Please write for a free catalog.
This book may be purchased from the publisher.
Please include $2.95 postage.
Try your bookstore first.

We are interested in hearing from authors
with book ideas on related subjects.

DEDICATED

to the vision, ingenuity and determination of the pioneers of West Coast logging.

FOREWORD

"SOMEDAY," said Big Fred Hewitt, "these pictures will show how the boys used to do it." How right he was. Big Fred was a saloon keeper, the owner of the famous Humboldt in Aberdeen, but he had vision that went beyond whiskey and dollars packed in a safe deposit box. He knew the time would come when the Big Woods was only a fog-blurred name and the cry "Logs! More Logs!" fast fading down the Skid Road of Memory.

Of course Fred Hewitt was not the only fellow in those early days who knew future generations would want to know what went on beyond the jumping donkeys and clacking blocks. The big point was he not only had a gallery of photographs proudly displayed but he was one who encouraged photographers to record the things going on. Enough of them did and enough loggers and lumbermen paid the freight to build a backlog of pictures that cover every operation in every phase of West Coast logging. Many of these records have been lost but many long buried have come to light.

In the foreword of "This Was Logging", which was limited to the photographic work of the master—Darius Kinsey, this writer said no one had attempted to round up enough good photographs and early data to create a book acceptable to publisher and public . . . that it would be a colossal task to do so and the result might be a hodge podge. However, the acceptance of that book "changed the picture" as it were. "This Was Logging" had the effect of awakening many collectors, large and small, to the fact that their prints and negatives were potentially valuable assets to the writer. It has been very gratifying to receive many voluntary offers of photographs from firms and individuals and a generally wholesome willingness to cooperate in any way possible has prevailed.

So the task of getting material for "Glory Days of Logging" was not a colossal task. After contacting many known owners of logging prints and exhibits, the writer went hither and yon into bush and by-street to make selections. As time progressed other people made photos and data available and as one lead led to another the volume increased to the point where late offers had to be regretfully declined. In most cases photographic copies were made of the originals which were then returned to the owners. A further sifting was done when engravings were ordered. The resulting quality, perhaps not as consistently fine as Darius Kinsey's prints, is nevertheless above average. The fine screen copper engravings made by the Artcraft Engraving and Electrotype Company of Seattle bring out the best character of the photos.

The extensive research made for written material in this volume was also pleasant work because of the willing assistance of librarians, universities, and lumber firm officials. These primary sources are indicated on various pages but particular thanks are due Eric Druce, John Collins, Douglas P. Taylor of British Columbia Forest Service; Willard Ireland, librarian and archivist, Province of British Columbia; Professor F. M. Knapp, forestry faculty, University of British Columbia; David Jeremiason, Booth Logging Company, Vancouver, British Columbia; G. H. Wellburn, MacMillan-Bloedel Lumber Company, Deerholme, British Columbia; Robert E. Swanson, Chief Inspector of Railways, Province of British Columbia; Lyle Stow, Mt. Vernon, Washington; Elmer Critchfield and Ernest Cogburn, Critchfield Logging Company, Port Angeles, Washington; G. M. Rhebeck and J. K. Lewis, Rayonier, Inc., Hoquiam, Washington; Ronald Todd, reference librarian, University of Washington; Marion Reynolds and Mildred Hill, Washington State Library, Olympia, Washington; Martin Schmidt and Inez Fortt, Oregon Collection, University of Oregon; Mrs. Hazel Mills, Multnomah County Library, Portland, Oregon; Herbert J. Cox, Lumberman's Buying Service, Eugene, Oregon; A. A. Lausmann, Kogap Lumber Industries, Medford, Oregon; Alfred D. Collier, curator, Collier State Park Logging Museum, Chiloquin, Oregon; C. Russell Johnson, B. J. Vaughn, Alder Thurman, Union Lumber Company, Fort Bragg, California; Harold G. Schutt, Tulare County Historical Society; W. E. Steurwald, U. S. Forest Service, Missoula, Montana; Charles H. Scribner, St. Maries, Idaho.

Ralph W. Andrews

Glory Days
of
Logging

Frontispiece

LOG TRUCK ROAD averaged 10 percent grade for 5 miles—Lake Grandy Timber Co. (Darius Kinsey Photo from Jesse E. Ebert Collection)

BRITISH COLUMBIA

It smelled like new country. The air was a little heavy that first morning, pungent with an aroma you figured came from a mixture of salt water and wood smoke. Both were all around you, around the mammoth sawmills and shingle mills on Burrard Inlet, False Creek and the Fraser.

There was nothing new about those mills except the size of them. Everything was so big—fir and spruce logs the like of which you'd never seen climbing up a jack chain. And the great booms alongside with more men walking them than you'd ever seen inside a Wisconsin sawmill—almost. But that was it—everything was steamed up on a grand scale and the acrid green fir smell accentuated the activity.

This was the Vancouver you had left Minneapolis to see and everywhere you were getting your money's worth. It was new country, still pioneer country with rough, uncouth edges and growing pains showing up here and there. You noted differences in the people too — fewer Scandinavians, more Scotch and Irish, the odd French Canadian and turbaned East India man and plenty of Chinamen. You noticed a different speech. It was Canadian.

You had seen skidroads, bigger and more sordid — Washington Avenue in Minneapolis, Chicago's West Madison—but they were not the rendezvous of loggers or seamen and had acquired a certain permanence. Cordova Street, Main and the others here still had the shifting uncertainty of places undergoing change. This skidroad was the real McCoy, not gloomy and dark holes for the hopeless and homeless.

(Opposite) **LOG TRAIN IN BIG CREEK GORGE** near Knappa.
(Weister Co. photo, Oregon Collection, University of Oregon)

FIRST SAWMILL ON SHAWNIGAN LAKE, Vancouver Island, B. C. W. E. Losee, builder and operator of mill, shown in boat. (Photo courtesy Gerry Wellburn)

You saw all this with the eye of a man who had seen the Middle West, who had worked in a Chicago saw factory and sawmills in Antigo and Rhinelander, who had sold saws in camps, veneer mills and woodworking plants. He wanted no more of crowded industrial living. He was looking for new fields to stretch out in. This looked like it. This was Vancouver—new country with a British tang to it.

Out beyond the city, up the northern mainland and across the Strait, was that fabulous reach of timber you had come to see. The loggers, in "to get their teeth fixed" or see if Queenie was still true, told you about Powell River, Alert Bay, Minstrel Island, Knight Inlet and all the vast forests and logging shows standing on end.

They told you about the finest stand of fir ever grown at Alberni, some of it merchantable timber when the Battle of Hastings was being fought in 1066. That you had to see (and you eventually did). Those tough Scotch-Irish buckers and chokermen told you how Davis rafts were built to go to sea and about the big high lead outfits and hand logging at the other end of the scale. It was a fairy tale of outlandish proportions, something beyond your imagination. But you'd asked for it and here it was.

You felt as a young fellow by the name of Mart Grainger must have when he came out from England back in 1906, going to work in the woods. He went along Cordova Street wide-eyed at the faller's axes, swamper's axes—single-bitted, double-bitted. Fascinated with the screw jacks and pump jacks, wedges, sledge hammers, crosscut saws and boom stick augers.

Grainger walked down Dupont Street where the shops gave way to saloons and the labor signs read: "50 axe men wanted at Alberni. Sooke — Rigging Slinger $4. Buckers $3½ — Swampers $3 Alert Bay." He elbowed into bars or sat with loggers who were "blowing her in" or waiting for boats north. They talked about Teddy Roosevelt, the Trusts, Socialism and the Yellow Menace.

He saw Bob Doherty punching men out of the Eureka bar and he got acquainted with Jimmy Ross' place where a logger who might have spent half his stake and if Jimmy knew him, got a ten spot handout. And then when that was gone and the logger had to ship out again, Jimmy Ross would sign him on for a northern camp. And Grainger saw Wallace Campbell standing drinks to men in the Eureka with one hand and hiring them for the day they sobered up, with the other.

This young fellow from England, Mart Grainger, worked in the British Columbia timber for two years. His full name was M. Allerdale Grainger and when he went back home he recorded his experiences in a book published in 1908—"Woodsmen of The West." The book, long since out of print, contains some crystal clear word pictures of men and conditions of the time. His characterization of his employer, Carter, as the new owner of a steam donkey, gives simple access to logging of the period. So now you go back a few years to Mr. Carter and the Big Woods as Grainger saw them.

"One of the great moments in Carter's life was that in which he paid the last installment owing to the sawmill and looked with proud eyes upon a donkey engine that was his very own. There, close by the beach, lay the great machine, worth, with all its gear, five thousand dollars. There, Carter could tell himself, was the fine object he had won by courage and by sheer hard work. There was the thing his earnings had created. Past earnings were no idle profit. There they were, in that donkey, in material form, working for him—helping him to get out logs and rise higher to Success.

"I make myself a picture, too, of an earlier moment in Carter's life — on the first morning when his donkey began its work. He sees smoke whirling up among the forest trees; he sees the donkey's smoke-stack above the rough shelter roof; the boiler, furnace, pistons beneath. And then the two great drums worked by the pistons, drums upon which are reeled the wire cables. And then the platform he himself has made, twenty feet by six in size, upon which boiler, engine, drums are firmly bolted: a platform that is a great sleigh resting upon huge wooden runners; hewn and framed together sound and solid.

"Watch Carter when the "donk" (his donkey!) has got up steam—its first steam; and when the rigging men (his rigging men!) drag out the wire rope to make a great circle through the woods. And when the circle is complete from one drum, round by where the cut logs are lying, back to the other drum; and when the active rigging slinger (his rigging slinger!) has hooked a log on to a point of the wire cable; and when the signaller (his signaller!) has pulled the wire telegraph and made the donkey toot . . . just think of Carter's feelings as the engineer jams over levers, opens up the throttle, sets the thudding, whirring donkey winding up the cable, and drags the first log into sight; out from the forest down to the beach; bump, bump! Think what this mastery over huge, heavy logs means to a man who has been used to coax them to tiny movements by patience and a puny jack-screw . . . and judge if Happiness and Carter met on that great day. (Continued on page 16)

BULLS IN BRITISH COLUMBIA. Leading team through T. B. Charleson's camp near Turner Mills in 1890. (Photo British Columbia Provincial Archives)

SKID ROAD ON PRESENT SITE OF VANCOUVER running from Marine Building location to Thurlow Street. Print made from original glass plate. (Photo British Columbia Provincial Archives)

The City of Vancouver is built on what was once fine timber land. Fir, spruce and cedar stood in thick stands, first affording Indians good cover for hunting and hiding, then furnishing the early British Columbia loggers good timber on flat land close to tidewater.

As you stand in Stanley Park (which was logged five times before 1889) you can almost hear the creaking of the yokes on the necks of the straining bulls, the rattle and clank of chains, the shout going up as a big log slews around on the skid-studded trail and noses into the brush. Then you stand aghast at the raucous bellow of the bullwhacker as he digs his goad into the oxens' rumps:

"Up Mack! Bright! Ham! By the jumped up devil that's in you, lean into it! Moony! Blackie! By God—I'm comin'!"

Vancouver's early loggers were Norwegians, Finns, Scots, English and Chinese, their lone tool the "Hudson Bay" axe, single-bitted with a curved heel and later the double-bitted axe. These were the swampers who cleared the brush for the skid roads; the skidders who cut short hemlock logs and embedded them crosswise; fallers; barkers, who limbed and peeled the logs; snipers who shaped the butt ends so they wouldn't hang up on the skids; doggers who drove in dogs and linked the logs together.

These loggers were also the greasers who followed the bull teams just ahead of the lead log and spread whale oil on the skids to let the peeled logs slide over easily and of course the strength of the pack —the bull punchers.

The skid roads ran from Point Grey area to all sides of the peninsula, from Roger's Camp into False Creek, from a camp near 25th and Trafalger to foot of Bayswater Street, from 29th and Balaclava to Jerry's Cove. W. H. Rowling had a log camp at 4th and Granville in 1891. There were several skidroads from Trout Lake and China Creek into Burrard Inlet and camps and roads in the Stanley Park district. The present streets of the city bear little relation to those early skidroads except Kingsway which follows the False Creek trail fairly closely.

BULLS IN BRITISH COLUMBIA. (Above) Team at landing. Note chains dogging logs together. (Below) Driving teams up skid road from Thurlow Island landing. (Photos British Columbia Provincial Archives)

"Carter's old donkey-engine was a mechanical chimera, and yet perhaps no worse than many others in the Western woods. The work it had to do was, of course, severe. The hauling of a blundering, lumbering log of huge size and enormous weight through all the obstacles and pitfalls of the woods; the sudden shivering shocks to the machine when the log jams behind a solid stump or rock and the hauling cable tautens with a vicious jolt; the jarring, whirring throb when the engineer hauls in the cable with a run to try to jerk the sullen log over some hindrance—all this puts a great strain upon the soundest engine. The strain of such work upon Carter's enfeebled rattle-trap was appalling. The whole mechanism would rock and quiver upon its heavy sleigh; its different parts would seem to sway and slew, each after their own manner; steam would squirt from every joint. The struggling monster within seemed always upon the very point of bursting from his fragile metal covering. In moments of momentary rest between the signals from the woods, the engineer would sprawl over his machine with swift intensity. Spanner in hand, he would keep tightening nuts that would keep loosening; it was a never-ending task. Hauling would often be interrupted, too, for more serious repairs. But still it was wonderful what the machinery would stand. One way or another the donkey did its work, and that was all that Carter cared.

BEAR CREEK FLUME AT ADAMS RIVER—Chase, B. C. (Photo British Columbia Provincial Archives)

FIRST STEAM DONKEY IN BRITISH COLUMBIA. Vertical spool, flat-geared donkey at Chemainus Camp in 1902. (Photo British Columbia Provincial Archives)

"Carter, you understand, does not belong to the class of ingenious-minded men. He is not skillful; he does not improvise ingenious makeshifts; he does not readily pick up new knowledge. When he bought his donkey, for example, he knew nothing about the care of machinery or the handling of engines, and he was slow to learn. So, for a time, he was obliged to depend upon hired engineers; to risk his precious donkey in the hands of men whose skill he had no means of estimating. But when he had gained a poor smattering of mechanical knowledge his rough self-confidence made him feel that smattering sufficient. Then Carter began to handle his donkey in his own way.

"Skilled artists — hook-tenders, rigging slingers, engineers — hated to work for a man who had never learned the ABC of classical methods. Carter did without such men. He went at every problem by the light of nature — "bald-headed," as the saying is—in furious attack. He would anchor out his wire cable around some tree, and make the donkey wind itself up mountain slopes, over rocks and stumps and windfall logs and all the obstacles of new-felled hillside forest. He would "jump the donk" aboard a raft from off the beach and tow it here and there along the coast. He did the things that skilled donkeymen can do. He handled his donkey in a stupid, clumsy fashion; muddling with it for want of skill, experience, and training; refusing assistance or advice from men who could have helped him. And yet he made that donkey go, in the end, where he willed it should go. He made it do his botching work, and made that botching work most profitable. He had no awe of his donkey, that great, awkward mechanism, nor of its ailments. He used it as in earlier days he may have used a wheelbarrow, as a thing that could be trundled anywhere, with freedom. But he had some heart-gripping accidents. Once, I have heard, some stupidity of his allowed the donk to slide downhill and drop into the sea. Bill was dispatched with the steamer to seek assistance, to ask some other logger to bring a donkey and, with it, drag the sunken machine to land. But no owner would expose his donkey to risk from wind and sea for Carter's sake. At last old Cap Cohoon came with all his men, bringing blocks and tackle and wire cables. His crowd and Carter's men between them drew the donkey upright in the water. There it stayed until the time of the "big-run-outs," when the tides go very low. Carter lit a fire in the furnace one night, got up steam, tied the cable to a tree-stump near the shore, and made the donkey wind itself up the beach—just ahead of the rising tide. And so he regained his donkey, his fortune. But the machinery was no better for the adventure.

"Another time when Carter was moving camp from Broughton Island down to Gilchrist Bay disaster hovered over him for two whole days. His steamboat, the Ima Hogg, was towing the whole outfit down the channel; towing the raft on which the donkey stood, the bunk-house raft, the cook-house raft, the office raft—a floating village. Heavy blocks, tackle of all description, huge hooks, wire cables, logging tools, boom chains, stores—every single thing that Carter owned (except his timber leases) was on those rafts. Suddenly, in mid-channel, the Ima Hogg lost her propeller!

"There were Carter and his hard-earned wealth left drifting at random, at the mercy of the tides. Wind might be expected at any moment in that neighborhood. Wind and sea would shatter his rafts and buildings, would send his donkey and his steamboat to the bottom, after pounding them against the steep, jagged, rocky shores. . . . I have heard that Carter worked for forty-

FAMOUS HASTINGS MILL—Vancouver. Panorama of historic sawmill in 1913. (Photo Leonard Frank Collection, Vancouver, B. C.)

eight hours fixing things aboard the rafts; and that having done his best, he went to bed and slept. He and his men were found asleep by Bill, who had gone with other men in a rowboat to search the channels for a tug; who had after two days found one; and who had returned with it in time to save the rafts; and Carter had had one of the most marvelous escapes from ruin that I have ever heard of in the logging country.

"You might think that such an incident would have shaken Carter's nerve and made him shy of risking his donkey upon sea-journeys. But barely six months later he hazarded his whole wealth upon a venture bristling with risks, the great venture of his life that brought him to the pinnacle of his success. It came about through the agency of a man named Billy Hewlitt.

"About this time, it should be said, logs were going up in price rapidly, and speculators had begun to realize that the forests suitable for logging (by existing methods) were limited in area and might soon be passing into private ownership. There arose, therefore, a great scramble to stake good timber leases. Parties of men explored the coasts everywhere for timber that was worth the staking; and other men in stores and bar-rooms and offices in Vancouver City gambled in leases of the timber that was staked. It was boom-time. Now Billy Hewlitt was a "timber-cruiser" — a man who sought for forest timber, to stake it; and Billy was hard up. For he was a man too hopeful, too enterprising. He had taken up timber leases in the most distant, unheard-of places. Dealers would not buy them—would not even send an expert to inspect them, so far away were they. The rent Billy had to pay the Government per square mile of lease was sucking his pockets dry. Things were thus going badly with him when, one day, he rowed his boat in to Gilchrist Bay and stayed at Carter's camp, storm-bound. Now Carter, working in his camp, had sniffed the smell of boom-time from afar. He had been cruelly torn in soul. He was making such good money, he was hurrying logs into the sea with

LOGGING AT SOOKE—Mayo Lumber Co. (Photo British Columbia Provincial Archives)

WHISKER RITUAL IN CAMP. Sunday morning stump-and-barrel ceremony at Chemainus in 1902. With washing on the line, congregation is solemn as head whisker faller makes undercut. Paper in victim's hands is assumed to be prayer book, not dinner menu. Note mug and brush on stump in wood smoke. (Photo British Columbia Provincial Archives)

such intense desire to profit by high prices, that he dared not leave his camp. Yet his gambling nature longed passionately to take a hand in the fascinating game of staking timber of which he had heard such glowing accounts from recent winners. So when Billy Hewlitt spent an evening at the camp, and talked big about the wonderful good timber he had for sale, and backed his words with the logic of two bottles of whisky that he brought up from his boat, Carter's heart took fire. He ordered Bill to load the steamboat up with fir-bark and get her ready for a cruise next morning. Then he and Billy Hewlitt steamed away among the channels, on a tour of inspection.

"Carter bought all these leases; dirt cheap, of course, for Billy was no match for him in cold business duels. And thus it was that Carter came to own, among other claims, the two square miles of timber at the head of Coola Inlet. When the cruise was over and he was back at logging work his thoughts would often dwell upon those two square miles. The sea-front timber was good.

"There is talk enough of Coola Inlet elsewhere in this book, and after reading it you may have some respect for Carter's courage in the great enterprise he now undertook, after deep thought upon the recent purchase of Billy Hewlitt's leases. Remember that Carter, after all, was a "small man"—a man in a small way of business. His little capital was new-made; he might have

RAILROAD LINE AT PORT NEVILLE—
Mainland Cedar Co. (Photo British Columbia Provincial Archives)

LOG CHUTE — THURLOW ISLAND.
(Photo British Columbia Provincial Archives)

given way to reasonable fears of losing it; he might have made a cautious choice of safe investment for it; he might have kept on working as he was doing, under moderate risks. He might have known that he was forty-six years old, and getting older after a hard life. Instead of that, by one Napoleonic stroke, Carter decided to take a risk that would have daunted a young man with five times his capital, that would have made a rich speculative company think twice. He decided to shift his camp and donkey to log the timber at the head of Coola Inlet, up among the feet of mountains, sixty miles of storm-swept water away from anywhere."

QUEEN CHARLOTTE INDIANS FALLING BIG SITKA SPRUCE. (Photo British Columbia Provincial Archives)

NEW "SHOES" FOR DONKEY. Skids for steam donkeys called for skilled axeman. Capilano Timber Co. (Photo British Columbia Provincial Archives)

YARDING ENGINE AND LOKEY AT CHEMAINUS IN 1901. Croft and Angus were sawmillers here in the early '80s. (Photo British Columbia Provincial Archives)

COOK'S QUARTERS—SHAWNIGAN LAKE—1908. Mrs. Burdess, cook for Shawnigan Lake Lumber Co., British Columbia, with daughter Mary and son Jim, now with British Columbia Forest Service, Duncan. (Photo Gerry Wellburn)

In 1912, the company was host to the royal party of Duke and Duchess of Connaught, then Governor General of Canada, and Princess Patricia of Connaught. The following day, Mrs. Elford, wife of the company manager, Theophilus Elford, received the following note bearing the Connaught coat-of-arms:

Dear Mrs. Elford:

I am desired by Her Royal Highness to send you the enclosed pin which she hopes you will accept as a small memento of today's visit to a lumber camp. Would you kindly give Mrs. Burdess, who served such a good lunch, the enclosed brooch from Her Royal Highness and ask her to accept it as a small souvenir.　　　　　　　　　　　　　　　Yours sincerely
　　　　　　　　　　　　　　　　　　T. H. Rivers Bulkely, Equery-in-waiting

FORWARD BAY CAMP—CRACROFT ISLAND—1947. Planked road of A & M Logging Co. Ltd. (Division of Huntting-Merritt Ltd.) shown running from timber to log dump and around point at left. Entire operation destroyed by fire in 1952. Crews were driven off island and brought to Beach Camp of Canadian Forest Products Ltd. at Englewood on Vancouver Island. (Photo Leonard Frank Collection, Vancouver, B. C.)

Sad Story
of the
CHINA TIES

Gerry H. Wellburn, resident manager Shawnigan Division, MacMillan-Bloedel, Ltd., can now smile at the China Tie incident, but in the depression year, 1933, it was tragedy. Ten cars of fir railroad ties cut by G. E. Wellburn Lumber Co. had been sold by H. R. MacMillan Export Co. for delivery to China. Cars were loaded with every stick checked against the mighty important order, and the train pulled out for Victoria and the waiting China ship. Six miles from the mill the axle of one car broke and derailed the train. Fighting desperation and discouragement, all hands fell to and with the help of Canadian National Railway section men, managed to use some of the ties to pry and jack the engine and cars back on the track. The order had to be delivered and the ship was about ready to leave. Other crews, aided by hastily recruited farmers and boys, and a hand pumped speeder, fanned out along the track and picked up every tie. Then the tally man reported they were exactly ten ties short.

No trouble to get ten ties? And what difference would ten ties make? Plenty of trouble—and these were hard times. To save fuel and use all available men on the wreck, Wellburn had ordered the mill shut down. And because Wellburn was a hard working young man of principle, and not to be hurried into one more error, steam had to be made, the mill started up and ten ties manufactured.

They were and the Chinese got full value for their gold.

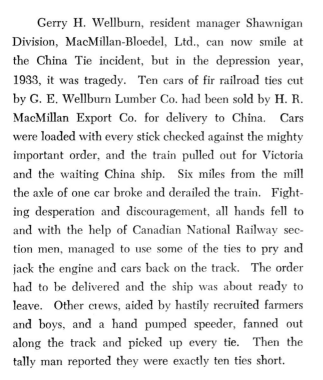

(Top) The mill that cut the China Ties. G. E. Wellburn Lumber Co. circular mill at Deerholme, B. C., cutting 20,000 feet a day. (Center) Guilty axle and spilled ties. (Below) Three of the ten ties that fell by the wayside in depression accident. (Photo courtesy Gerry Wellburn)

KING OF THE BULL COOKS

By
STEWART H. HOLBROOK

Logging's most able writing man, Stewart Holbrook, progressed from British Columbia camps to lumber journal humor and editing that stalwart little organ of the Loyal Legion of Loggers and Lumbermen—the 4 L Lumber News in Portland. His classic of the woods, 'Holy Old Mackinaw," heads a long list of books bringing him up to the current epic "Columbia."

Early in the summer of 1920 I hove into a logging camp in northern British Columbia that was two days by boat from Vancouver. The boat paid little attention to the line's printed schedule and it was late at night when we docked at Deep Bay. As I walked down the gangplank to a log float that was lighted by a dim kerosene lantern, a squat, powerful figure approached me from out of the gloom. It was a man some five feet tall and about as wide across. "You th' new cheater?" he asked. I told him yes, that I was the new timekeeper and scaler.

"That's what I thought," he said. "Foller me." He led the way down a plank walk, past bunkhouses from which came the noise of sleeping loggers, to the camp office. "She's a great layout here," said my guide loudly, "but you gotta be good to last. She's highball."

I was quite sure this fellow was only the bull cook, the camp varlet, and I didn't think a bull cook ought to talk that way to so important a man as a timekeeper and scaler. So I thought I'd put him in his place with some neat sarcasm. "Are you the camp push?" I asked. "No," he said, "I ain't the camp push. I'm the bull cook." Then, after a brief pause: "I'm the bull cook, the King of the Bull Cooks. I'm Okay Fuller."

The voice had the tone of assured authority, and it also carried considerable pride. I looked Okay Fuller over. He was, as I said, about as wide as he was tall. He was somewhere in his late sixties. He had a shaggy beard, grizzled but not yet white, of the General Grant mode, and his face and neck were tanned and wrinkled like an old pair of logger's tin pants. He wore congress shoes, the first I had seen in a long time. His galluses were stupendous, so wide and ruggedly built that they gave the appearance of holding him down into his pants rather than holding them up around him. On the gallus buckles, instead of the usual "Firemen's" it said "Hercules." ("Ordered special from the Hudson's Bay Company," Okay told me.) The top of his pants, I figured, came just below his breastbone.

"Got any drinkin'-likker?" he asked as he lighted a lamp in what was to be my room off the camp office. I told him I hadn't.

"That's no manner of way to come into camp," he said severely. "It usta be that we had men in the woods." I apologized that I had started from Vancouver with a quart but that it had been used up on the long trip North. "Maybe you'd like a snort, then?" he sort of asked. I said I would.

Okay went back of the office counter and fumbled under it a moment. When he came up he had a small package in his hand.

Without more ado he tore the carton away from a four-ounce bottle, yanked out the cork, and poured about a third of the contents into a tin dipper of water. The water turned as white as milk. He passed the dipper to me without a word. "What is it," I asked, "jakey?"

"No, it ain't jakey," Okay said. "It's painkiller. Drink hearty. Good for the bile and eases the kidneys." I drank deep and the stuff hardly hit my stomach before I could feel it in my ankles. I hurriedly gulped down a chaser of water. Okay took the dipper and turned the remaining two-thirds of the painkiller into it, tossed in a bit of water, and drank it, unflinching. "Painkiller," he announced as though addressing a meeting of the Canadian Medical Association, "is a sovereign remedy against the night chills of the forest. Most especial in the damp an' humid forests of th' Pacific slope."

We talked a while and Okay told me what sort of a camp it was. "This is a mighty fine place," he said, as if he were trying to sell it to me. "You'll like it. Palace in th' woods, that's what I call it. Palace of all th' sultans. Lux'ry such as known only to princes, popes an' po-ten-tates. And me, I'm th' grand eunick. Make yourself to home." Then he went out, slamming the door behind him.

While undressing I looked at the label on the painkiller bottle. It claimed to be of great thaumaturgic power, good for man and beast, internally and externally, and was graced with a steel engraved portrait of what appeared to be an eminent savant of old-school medicine. I slept very well.

HOLBROOK AND "THE GREAT DANE" AT GRASSY BAY. Two logging characters in search of an author in 1922. On right, with corncob pipe and tin pants, Stewart Holbrook who much later was to write the lumber industry classic "Holy Old Mackinaw" and Murphy, "The Great Dane", noted skid roader from Pender Harbor to Loughborough Inlet for thirty years. (Photo Stewart H. Holbrook Collection)

The camp at Deep Bay, I found, was just about average, but Okay Fuller was a bull cook who was anything but average. He liked often to announce that he was king of bull cooks and there is no doubt in my mind that he earned the title, hands down. It should be explained, possibly, that the bull cook of a logging camp doesn't do any cooking. The title is ironic. He feeds the pigs, sweeps out the bunkhouses, cuts and brings in the fuel, washes the lamps, and generally acts as a chore boy around camp. Bull cooks usually are superannuated loggers, and they are often old soaks, bleary of eye and full of locomotor ataxia, with no spirit. Okay Fuller was none of these.

For one thing there was the way Okay handled his homely duties. He had some two hundred kerosene lamps that must be washed every few days, and filled every other day. Okay called science to his aid. Instead of the long process of wiping the washed chimneys with rag or newspaper, which would have taken hours, he first dipped the chimney into a bucket of clear hot water. After this operation he stood the chimneys on a warm stove. The steam thus generated left the glass as clear as crystal, and the job was done in a moment.

In the matter of splitting large and tough chunks of fir to feed the ravenous box stoves and cookhouse ranges, Okay approached genius. He had the blacksmith pound out a sort of thick metal cartridge with a wedge-shaped end which had a hole in it. He would fill the cartridge chamber with black powder, drive the wedge up to its hilt into the wood block, and touch off a short fuse. The result was a neat quartering of the block. The accompanying noise and smoke, and the smell of powder, pleased Okay immensely. He always said it reminded him of the time he served in the Reil Rebellion in Manitoba and Saskatchewan, Canada's tin-pot "civil war" of the eighties. One time the illusion of battle went further. His own belly well filled with potent painkiller, he tamped an extra heavy charge into the wedge. The explosion not only blew all hell out of the block, but the gadget burst and part of the metal went through Okay's hat. He had the smithy make a new and larger one. "Make her strong as the Old Rugged Cross," he ordered.

And Okay knew how to deal with bedbugs. They got pretty thick in camp that winter. We tried sprinkling the bed clothes with kerosene, which made our legs raw and sore but didn't seem to have much effect on the pests. One Sunday Okay prodded the engineer into steaming up the locomotive. They rigged up a piece of hose to the condenser, backed the engine up to one bunkhouse after

STEAM TRACTOR AT GILLE'S CAMP. (Photo British Columbia Provincial Archives.)

another, and packed them full of steam. It was three or four months before we saw a bug again. In appreciation of this noble work we all chipped in and had a gallon of Demerara rum sent up from Vancouver for Okay, and helped him drink it.

But in spite of his scientific and mechanical interest in things that applied to his own work, Okay Fuller was a hidebound Tory when it came to anything he considered newfangled. Once he evinced a friendly interest when somebody told him a cock-and-bull story about how farmers down in the Fraser River Valley were making fine potable alcohol simply by pouring hard cider into a cream separator and turning the crank; but all other inventions, he said, were leading us to pot.

Although there had never been an automobile or airplane within three hundred miles of camp, Okay was constantly bellyaching about them as works of Satan. The single-shot rifle he had carried in the Reil campaign he held to be a far deadlier and more accurate weapon than the machine gun. He had read something about what he called "the wireless radius (sic) machine" and was sure that no good would come of it. When he spoke of cigarettes he fairly frothed.

One day the weekly boat brought a large packing case for the camp. It proved to be a small electric light plant, and with it came an electrician to install the machine and wire the bunkhouses. For days while the work was going forward Okay muttered and seethed. He told how a house owned by his uncle had been struck by lightning the next day after electric lights had been installed. Justifiable act of God, he termed it. His appetite fell off and he took to guzzling more painkiller and prune juice—which he fermented in his shack—than usual. After he learned how to run the machine I think he rather liked it—that is, privately—as well as the fact that he didn't have to care for two hundred lamps every day or so. But he always claimed that "th' rays of 'lectricity is injurus to th' nekked eye, as well as havin' a bad effec' on the brains."

During that winter a number of small fires occurred in the bunkhouses, caused by newly arrived drunks who went to sleep with lighted cigarettes, or by loggers feeding too much "fat" fir into the stoves. And the grading foreman had an unsocial practice of keeping a few cans of dynamite caps under his bunk. Taken all together, the place wasn't so good as an insurance risk. So Paul, the camp foreman, ordered a fire extinguisher for each bunkhouse.

Okay Fuller, of course, was loud in his disapproval of the extinguishers. Every time his eye happened to light on one of the small, bright red objects hanging from a bunkhouse wall, he damned it from the end of its handle to the nozzle. He had, he said, been chief of a volunteer fire department back in Ontario; he was thus an authority and he knew, by God, that no fire extinguisher was so good

POLE LOADER AT SHAWNIGAN Lake Lumber Company in 1903. Note concave wheels on log rails. (Photo courtesy Robert E. Swanson)

as water. He went even further, and in a somewhat obscene comparison declared that one healthy poodle dog, male, could **lay** more fires than all the goddam so-and-so extinguishers put together. "Them red things," he told anyone who would listen, "are trumpery."

It went on this way for months, with Okay harping about the extinguishers. Finally, the foreman got tired of it. "I'll show that old devil what these things will do to a fire," Paul told me one Sunday. He made a pile of packing boxes and excelsior in front of the camp office, and at dinner that noon he invited the entire crew to be present at two o'clock to witness "the magic of Little Pluvius," a phrase he had retained from literature on the subject.

The gang was on hand—Swedes, Finns, Bohunks, and a handful of English-speaking loggers. Paul lighted the pile of rubbish and let it get to blazing good. "Now," he shouted with a flourish of the Little Pluvius, "now you shall see how this small object in my hand will extinguish Man's greatest friend and at the same time Man's greatest enemy."

Paul stepped close to the fire and turned the handle. A small powerful stream of liquid shot forth into the flames. The fire leaped upward all of ten feet. It was so hot Paul had to step back. He shot another round into the fire. It roared louder and higher. Paul's eyes bulged out and there was sweat on his brow. Back of me, in the crowd, I heard Okay Fuller state quite audibly that he had "knowed all along that that Pluto dingus wouldn't put out the fire on the end of a Swede match." He seemed immensely pleased about it.

The fire roared and grew mightily. The audience was moving back a bit. "Get some water," it was Okay's voice I heard. "Get some water," he shouted, "or the whol' jeasley camp'll burn up!" I noted that he already had a pail at his feet. He doused it into the fire, which slackened perceptibly. Okay threw on another pailful and the fire died down to smoke and embers. Okay turned to the audience. "My frens," he said, and it sounded for all the world like a speech for votes. "My frens, you can see what that Pluto trumpery amounts to . . . I thank the good Lord I happened to have a pail of water handy so as to lay low the flames that would have burned us out and made us paupers."

This resounding statement, coupled with the obvious fact that the extinguisher had not put out the fire and the pail of water had, took immediate effect on the crew. Swedes muttered as they tamped a fresh r'ar of snuff into their lips. One of the Hunkies, a small fellow but a most potent orator, made a speech that didn't sound good to Paul, even though he couldn't understand a word. The lone French-Canadian, one Alex Croutier, said he had heard of a man killed when a fire extinguisher exploded.

Later, in the privacy of the camp office, Paul and I investigated the liquid left in the extinguisher; it was pure gasoline. Paul charged Okay with doing the job; but there was no proof, and Okay would admit nothing except that God had taken a hand in the matter to show that His water was better than man-concocted chemicals.

The sun was getting to feel warmer. One day I sat on a block in the camp yard where Okay was cutting fuel for the cookhouse. The pungent smell of new sawdust told me that the sap was going

SPRUCE BOOM—NORTH VANCOUVER—1947. View of fine spruce at booming grounds of Evans Products Ltd. (Photo British Columbia Provincial Archives)

up the trees. The night before I had heard two cougars snarling and yelling, up on the mountain back of camp—mating cats. Birds were chippering around now. The brook had come to life; I could hear it gurgle over the stones. I remember I thought I would write a poem about it.

Okay had just left the woodyard, wheeling a barrow of fuel into the kitchen. There was a scream, then a wild yell from that direction. I hurried over as fast as I could, but it was all over when I got there. The cookee, or cook's helper, had discovered his wife, who did the dishwashing, in a condition with the boss cook which barristers in Canada describe as adultery. The husband-cookee had thereupon picked up a meat cleaver and started for the couple, shouting that he would cut off his wife's ears and commit mayhem on another part of the cook. Just then Okay happened in with his load of wood. Sensing the situation at once, he picked a hefty stick of fir from his pile and whaled the big cleaver-man over the head, laying him cold and quiet under the sink.

"Hated to do it," Okay told me afterward, "on account adult'ry being something that'll happen with every change of the moon and is mos' likely when the sap's in the wood."

Paul had to fire the entire cookhouse crew. Logging-camp ethics required it. And the incident had an effect on Okay, too. He got very restless. His eyes had a faraway look. Twice he forgot to feed the pigs. And next boat-day he came into the office wearing a rusty old derby hat, a stiff-front white shirt and a rather pretty hook-on tie. "Boy," he said, "I guess I'll go down to Vancouver for a few days. I want to have some work did on my teeth." This, of course, is the standard and classic excuse of the logger who wants to go where he can get things that come in bottles and corsets. I knew Okay Fuller didn't have a tooth in his head, and for the fun of it I told him so.

"Well," he said, with what he meant to be a sensual leer, 'maybe I ain't got no teeth, but one of my gooms is botherin' me."

Okay went down on the boat and I never saw him again. That was long ago. It may be by now that he has gone to the bull cooks' heaven where the lamps are self-fillers, the wood grows exactly the right size to fit a box stove, the bunkhouses are swept each day by perfumed zephyrs, and 110-proof painkiller oozes out of the trees. Anyway, if Okay Fuller has gone there, I am sure he has the situation well in hand.

UNLOADING POLES AT JEUNE LANDING. W. F. Gibson & Son dumped poles from truck with simple hoist and gin pole. (Photo British Columbia Provincial Archives)

DUPLEX 4 ON PLANKED ROAD at Cracroft Island, B. C., in 1929. (Photo Leonard Frank Collection, Vancouver, B. C.)

THE BALLAD OF THE SOILED SNOWFLAKE

From "Bunkhouse Ballads"

By ROBERT E. SWANSON

"THEY HAD NO POET — AND THEY DIED"

As a youth of Vancouver Island's East Wellington, Bob Swanson went into the woods pulling whistle for John Coburn in 1919. He became in turn sawmill engineer, chief engineer, civil engineer and mechanical superintendent at Victoria Lumber Company in Chemainus.

Since 1940 he has been inspector of railways for British Columbia—borrowed for the war years by the Crown in the airplane spruce program. His personal friendship with Robert W. Service may or may not have helped develop his aptitude for writing verse, the fact remaining that Robert E. Swanson is noted as Canada's rhymster of the woods. The following is this author's selection of one of his best, a second appearing on another page.

*The Madam stood in her parlor when a knock
     was heard on the door,*
*Her fairies then gathered around her to display
     their stock and store.*
*She peered through the grill in the panel like a
     panther stalks a deer,*
*And with a quick respond to the cute little blond
     she whispered in her ear:*

"He's fresh in from the jungles, dear, with a great big
     roll of hay,
So stick right close beside him, and make that sucker
     pay.
I had my spotter down last night to watch the boats
     arrive,
And my taxi driver picked him up in an east-end boot-
     leg dive.
He'll be my guest while you get dressed in your finest
     evening frock;
His tonsils anoint in a cocktail joint, but bank his roll
     in your sock.
Offer your charms to lure him — make sure of your
     feminine wit;
But get his jack, and then come back. It's a fifty-
     fifty split."

"Hello, dis place!" said Mickey O'Shea, as the Madame
     ushered him in.
"I'll quaff me some of your good old rum 'cuz I know
     you're drinkin' gin.
Here's to the ladies — God bless 'em, and here's to the
     rum — drink her down;
Skol! Bottoms up; best o' luck; fill the cup, there's
     plenty more liquor in town.
Now, trot out your girls for my choosin' for my flesh
     is seared with the flame
That has burned in man since the world began (O, need
     I mention its name?)"

"Now, dearie," the Madame intruded, "I know you're
     a-rearin' to go.
That roll you pack, of hard-earned jack, is a mighty
     big wad of dough!
Just be advised by one who's wise to the tricks of the
     Huntress Clan:
Steer clear of the harlot, the woman in scarlet, she's
     out to fleece you, man.
So let me make you acquainted with the right kind of
     gal to acquire,
She's honest and true and she'll see you through, shake
     hands with Molly Macquire."

"What a woman! My God! What a woman," thought
     the man from the log-jammed streams.
I'd follow her track to Hell and back, she's the girl of
     my fanciful dreams.
. . . I've lain all alone in the darkness of the forest
     with boughs for a bed,
With the towering pines up above me, and the mur-
     muring wind overhead —
And I've heard her voice in that stillness, and she's
     come like a nymph ere the dawn,
To soothe my soul with her fondness. With the stars
     and the night she'd be gone.
And always I've fancied I'd seen her in those big, deep
     pools of blue
Where the cataract leaps in the river, I've heard her
     laughing, there, too.
And now, just to think she's beside me — O God, but
     it's hard to believe . . .
"Come, Honey, let's head for a night club," said the
     blonde little daughter of Eve.

          *   *   *   *

Ten thousand drums and a big brass band, queer ani-
     mals, purple and red,
Were climbing the walls with ten pound mauls and
     a-thumping them down on his head;
A circus of serpents performing in a bathtub full of
     champagne,
They were long, they were lean, they were purple and
     green and a-writhing around in his brain.
. . . A woman — a marvel of beauty, with a form like
     a sculptor's dream,
With rippling laughter in her eyes, like the moon on
     a mountain stream,
Was calling him close to her bosom, was enticing him
     to her embrace;
But always her image would vanish and the floor would
     wallop his face.
Now, was it a horrible nightmare, or the jims from the
     liquid fire?
Had the Madame not made him acquainted with a girl
     called Molly Macquire?
He seemed to remember a road-house where the twin-
     kling, bright lights shone.
Then, sudden, he ran through his pockets — his roll —
     My God! It was gone!!

TRAIN LOAD LOGS—QUEEN CHARLOTTE ISLANDS. (Photo British Columbia Provincial Archives)

How cold were the streets of the city, how barren, how friendless and bare.

How lightly the snowflakes fluttered, pure white, on that turbulant air;

How soon were they soiled in the gutter, their emblem of purity stained.

Like the maiden whose visage he'd conjured in the forest where solitude reigned.

He was heading right back to his little log shack on the north-bound boat that night.

How he spent the day it was hard to say, yet, he seemed to remember a fight.

Battered and bruised and badly used, and nursing a big black eye,

While many a dame of skidroad fame was waving her love good bye . . .

When the steward tapped on his shoulder and beckoned him on with a sign:'

"If you're Mickey O'Shea, then follow me. You're wanted in stateroom nine!"

It was Molly Macquire, his heart's desire, that opened the stateroom door,

And she said with a grin, "Come on right in, you're blocking the corridor.

You'll think I'm tricky, but listen, Mickey," said the blonde as she opened her purse,

"Last night on the spree we saw a J.P. and you took me for better or worse.

And here's the whole of your hard-earned roll that you gave me to hold for you,

And now I'm your wife you can bet your life I'll always be honest and true!"

There's a house in the city that's builded on the sins by the Devil ordained,

Where the snowflakes drift in the gutter, their emblem of purity stained;

There's a little log shack in the forest where Repentance has kindled a light,

And the snowdrifts gleam in the valley, untrodden, untarnished and white.

HAND LOGGING

Many British Columbia power loggers got their start on hand logging licenses from the Crown. The steep-sided islands with inlets opening into quiet sounds made it possible for one or two men to make a living or a stake using only hand tools—double-bitted axes and jacks. Some made fortunes.

It was hard, back-breaking work but most hand loggers liked the solitude and independent life. They felled the trees, bucked them in salt water and towed them with skiffs into a small boom which was power-towed to mills at Vancouver.

The work was dangerous and if badly hurt, a man working singly had no way to get word for help. Stumping shows, where logs would fall to a slope and on down to salt water, were comparatively easy and safe. But if a tree had to be run or worked part way down a slope, a man could easily lose a leg or a life. M. Allerdale Grainger, in his book "Woodsmen Of The West," gives a vivid picture of his boss Carter when he was hand logging.

" 'The first time I ever seed Carter,' Dan Macdonnel said to me one day, 'was in a camp on Puget Sound where I was blacksmith. Carter comed and worked in the camp—just the same Carter that he is now—a desperate man to work, surly, and wanting to do everything according to his own ideas; thinking he could handle any job whatever in the woods, and show men who had worked all their lives at that job the right way to do it, whereas he can't do no more than butt his way through after a fashion. He used to be a nuisance to work with unless a feller let him have all his own way. I know the boss at that camp had to hold himself in all the time, to keep from losing his temper and firing Carter. But he felt there was no sense in losing a good worker like him. That was why Carter was able to stay so long.

" 'Before he came to our camp Carter had put in a few weeks lying round Seattle; drunk most of the time, but still hearing a good deal of talk. He had come across some men that had been up among the islands and inlets on the B. C. coast. They told him there was a growing demand for logs on the Canadian side, and that men were able to go up north 'most anywhere and make good money hand-logging. Carter got bitten with the idea of going up there himself.

" 'He was always brooding over the proposition, and whenever he'd get the chance he'd talk to us boys about it: what a fine show there was for a couple of men to go to Alert Bay and hand-log somewhere round them parts; and what big money they could make; and how they would be their own boss. You bet it was just poison for Carter to be doing work for another man.

" 'Then Carter would pick on some man or other and try hard to get him to go north, in partnership. He was after me one time. Now I was sort of willing to make a trip up and give the hand-logging a trial; not that I knew the first thing about it, but from what I could hear a man would soon get used to the work. But you wouldn't have caught me going as Carter's partner. Being partners with him means obeying him and being his slave; a man of any independence couldn't stay with him five minutes. Carter's as pig-headed as they make them; and wicked. Everything's got to be done his way; your way is wrong, and he won't even listen to what you are going to propose; and he'll go against your interests, and against his own, and wreck his whole business rather than admit himself in the wrong. You can't begin

HAND LOGGER JACK-ING UP FIR. With the log half peeled, hand logger gets forward end up to maneuver it into sliding position. (Below) Hand logger's camp at Tekearn Arm. (Photos British Columbia Provincial Archives)

to argue with him; he flies off the handle soon as you open your mouth. I've no use for a man that goes on like that. Well, he couldn't persuade me to go with him, but he got hold of another feller, and soon after that Bill Allen made up his mind to join them. The three men saved up their wages for some time, and then they all quit the camp and went down to Seattle to take one of the Alaskan steamers that used to stop at Alert Bay, going north. Carter had two hundred dollars saved up, and Bill had about five hundred. The other feller got drunk and missed the steamer, and they never saw him again.' And that was Macdonald's story about Carter.

"At the time when Carter and Allen came north, Alert Bay was still the nearest jumping-off place for the Broughton Island district. There was a big store there full of all necessaries, for Indians and white fishermen and prospectors and trappers and such-like men. Twenty or thirty hand-loggers, I believe, also drew supplies from that store; and hand-logging tools could be bought there, at exorbitant prices.

"So you can imagine Carter and Allen engaged in buying an outfit, and paying high for it. First they would get an eighteen-foot rowboat, with a good sail. Then tools: two heavy jack-screws, a light ratchet screw, big seven-foot saws, axes, heavy chains for chaining logs together, and many other things. Then flour and beans and bacon and the like in neat fifty-pound sacks, sewn up with oil-cloth; and tobacco in boxes; and a good-sized sheet-iron stove (with an oven up the chimney); and lots of matches (in that wet country) in tins; and maybe ammunition and a rifle. I bet Carter bought no fancy canned stuff, nor canned meats, nor any such rubbish; but he would have done himself well in cream and milk and syrup and little things a man really needs. And before the outfit was all stacked on the wharf Carter would have spent some five or six hundred dollars, cash down.

"The talk at Alert Bay decided Carter and Allen to go search for a hand-logging proposition in the channels round Broughton Island. You can, if you

STEAM FUEL SPLITTER—ENGLEWOOD. (Photo British Columbia Provincial Archives)

LOADER AND TONGS with duplex loading engine at Sisters Creek, 1933. (Photo British Columbia Provincial Archives)

HOPE SAWMILLS LOG CHUTE AT BROOKMERE. (Photo British Columbia Provincial Archives)

like, picture the boat trips: the over-laden boat; sailing winds; head winds; rough water; wetted cargoes; long, weary hours of rowing; runs for shelter behind islands; camps made in the dark by exhausted men. No men would spare themselves less than these two; no weather except the really dangerous would stop them. There was some queer anecdote about their power of endurance that I wish I could remember. But all I know is that they brought their stuff to the north end of Gilford Island, and 'cached' it there, and started out with unencumbered boat to seek and choose a place where they should work.

"In those days good timber was plentiful — good timber, on sea-coast slopes, that could be felled and shot right down to water—hand-loggers' timber. The country bristled with opportunities, for loggers; opportunities that were the making of men who had the spirit to venture out and seize them— men like Carter, opportunities that were then new-born of changed conditions in the lumber trade. Bitter to the Westerner are the mistakes of caution.

KINGDOM IN HIS GRASP. Whistle punk at Beban Logging Co., Buckley Bay. (Photo British Columbia Forest Service)

"Many a man I have heard lament those days. 'Boys, oh boys!' one would say, 'why was we all so slow in coming to this country? We'd heard talk of it, and yet we held back: pessimists, that's what we are. Men like Carter got ahead of us: he had us all beaten. Why, anywhere round here all up the Inlets and round the islands there were the finest kinds of hand-logging shows. Why! the country hadn't been touched! There's men working today on places that have been hand-logged, and re-hand-logged and re-re-hand-logged since them days. . . .'

"So Carter and Allen had no need to cruise far around the shores of Broughton Island. They saw a boom or two hung out in little bays that opened from the channels; they received welcome at the cabins of the few hand-loggers already working there; but soon they rowed their boat past untouched forest slopes and knew that they had pushed ahead of the advance of man and human work. Everywhere their eyes were gladdened with the sight of timber handy to the beach; fine big cedars for the most part. Many trees they noticed, pointing sudden fingers, would drop right into water from the stumps when felled; a thought that made their hearts feel light. For 'stumpers' are the most profitable trees that hand-loggers can hope to get; they need so little time and work.

"So the two men looked eagerly for a small bay where wind and waves could not blow in with any violence; and this they had to choose most carefully, by observation of the signs of weather on the beach and on the trees; and by argument. What would the west wind do in summer? How would the north winds strike? Which way would the sou'-easter blow from off the mountains?

"They found a bay that seemed to them secure from wind and sea, that lay close to a fine stretch of cedar forest. The hillside, too, rose from the sea at the right sort of angle; neither too steep for men to climb, carrying their tools, nor too flat for logs to slide down easily. A little creek fell with pleasant noise over the steep rocky beach of their little bay, and the two men found just by its bank a small flat place on which to build their cabin. So they pitched a tent, near to the shore, and by that act secured (by logger's courtesy) their title to the bay and to the neighboring slopes. Then they made laborious rowboat trips, bringing their outfit up from where they had it hidden. That done, they set to work to make their camp. They did not build the ordinary log-house, cedar was so plentiful. Instead, they cut a cedar log into eight and twelve foot lengths, and split the straight-grained wood into planks with their axes; and made a house-frame out of poles, and sheathed the frame with their cedar planks. Then they put in a floor of rough-hewn slabs; and

OLD TIMER AT SHAWNIGAN LAKE. First locomotive used by Shawnigan Lake Lumber Co. (Photo D. B. Taylor, courtesy B. C. Forest Service)

A-FRAME LOGGING ON BRITISH COLUMBIA COAST. Mainline runs from block in peak of A to tailblock on stump in logging area, which is moved in arc as falling progresses. This photo is of Pacific Great Eastern log salvaging operation. (Photo British Columbia Provincial Archives)

HAULING SHORTLOGS WITH CAT. Winter operation at Adams River Lumber Co., Chase, B. C. (Photo British Columbia Provincial Archives)

RIDING THE SKIES. In attempts to eliminate building of logging roads, in difficult terrain, work faster than high lead and spare young growth, ingenuity produced the "sky-hook" method, used in British Columbia to a limited extent. Cable track on which car rides is strung between spar trees 800 to 1,000 feet apart, like old department store change carrier. Driver of "sky-hook" car takes it out to falling area, picks up his load of logs and brings them back to landing. Wheels are rubber-tired with steering mechanism and gears retractable for use on roads. (Photo Jesse Ebert)

fixed up bunks, and made a table, and set their cook-store and its stovepipe in position. Outside the house they cleared a little flat; and underneath a shed they set their grindstone; and made a stand where they could sharpen their big falling saws. Their camp was soon completed. Morning and evening blue smoke ascended from it, and marked its site against the mountain slope; and the sun shining sent a home-like gleam from the yellow roofs seawards through the foliage.

"Now the two men took their tools along the hillside to where tall, slender fir-trees stood. These they felled into the sea, and cut a long sixty-foot log from each. They bored holes through each end of every log and chained the logs one to the other. So they had a long chain of logs to stretch across the mouth of their little bay. Anchored firmly to the shore on either side, that floating line of logs would give them harbour for the logs they meant to cut: once placed inside, no log could wander off to sea. Their 'boom' (in loggers' speech) was 'hung'. They were now ready to start hand-logging.

" 'We worked right straight along when we were hand-logging; none of this here laying-off for rain or blank-blank laziness. We made big money,' was all I ever got from Carter concerning this period of his career. And yet romance lurks there. For the things men do in company are, after all, the easy things. What is so easy as to play one's part in charges on a battlefield; to join a crowd in doing certain work; to add one's little mite to the pile one's fellow-workers make in sight of one? I find it impressive, I feel how much it is above the reach of average men, when men go out alone, or two or three together, against Nature in its wilderness; and there achieve noteworthy things by strain and stress of sweaty labour, hard endurance, laborious ingenuity. They work there, their own conscience driving them, with no crowd of fellow-men to notice what they do. They have no helpful standards of conduct held before them; they are free to stand or fall by their own characters, that lack the supporting stays in which the morals of the citizen of towns live laced. And yet such men as Carter will 'work right straight along' in dismal wet discomfort, in far solitary work, handling with imperfect tools enormous weights and masses — at mercy of callous Nature—undismayed."

HEEL BOOM LOADING at Forward Bay, Cracroft Island. A & M Logging Co. Ltd. was burned out here in 1947. (Photo Leonard Frank Collection, Vancouver, B. C.)

TIMBER PORTRAITS

"Well," the old timer was saying, "I guess I better crawl in the straw before the old lady chucks it out in the barn. Who? Jump-Off Jack? Why sure I remember him. I ain't so old I forgot that looney. Say, I remember . . ."

Out of such treasure houses of memories come all the facts, plain and fancy, that make men into legends. Get enough loggers thinking of the past and into a talkative mood and you begin to get a picture of the "glory days."

British Columbia seems especially rich in buried treasure of this kind or is it because this country has a chronicler of note with the expansive ability to pull a rare bird out of the bush and clothe him in gay feathers?

Robert E. Swanson has published and recounted on radio dozens of sketches about loggers of early times. Here are some of them hewn to fit this specialized space.

STEAM POT PIONEER

When the Hastings Mill burner was a beacon for ships in 1885, Saul Reamie was woods foreman. He whipped oxen and horses up and down the skidroads until Vancouver Townsite was logged off and in 1899 took his teams to Rock Bay. Here he added a railroad and steampots—vertical engines with gypsy drum, then single-drum, horizontal-drum and twin engines. But he kept the line horse, Jerry, shod with caulks so he could walk on floating logs.

Saul set up Camp O on Thurlow Island for Hastings Mill in 1907, using a locomotive named Curley and a few single-drum donkeys. When manager Sandy McNair sent up a new 2-drum donkey on a float with two reels of line, Saul ranted at the stupidity of town dudes, moved the donkey two miles up the skidroad and left the haul back in camp because he didn't know what to do with it.

But he was a friend of his men, like Greasy Bob Graham and Whiskey Shannon and always tried to keep them from leaving camp. "They always got work and eats here," he said. "Them fellers made Hastings Mill." One Lea Curtis, a colored hook tender had a run in with sniper Sam Oleson. Chased

UNLOADING AT MENZIES BAY. Campbell River Timber Co. used this jill poke to unload trucks. (Photo British Columbia Forest Service)

"OLD CURLEY" IN 1894. Saul Reamie's famous lokey at Camp O, Thurlow Island. Bob Harvey, engineer, is in cab, Perry Des Brisay leaning against first log. (Photo courtesy Robert E. Swanson)

into camp on the double, Lea was about three jumps ahead of the Swede's axe and kept right on going, yelling back to Saul—"It's better to say—'There he goes than there he lies'!"

Saul befriended an old section hand who had worked for him for years and was left $20,000 when the man died. When Saul died at 68, the engine "Curley" was placed in Vancouver's Hastings Park as a monument to loggers.

8-DAY WILSON

Holder of the British Columbia short stake cup, 8-Day claimed he had no home, no Christian name, and was hatched behind a stump. He would hire out to work eight days only and when they were in, he "bunched her" and went back to get the rest of the keg.

8-Day's favorite stunt was to go into camp as a hook tender, lie around the first day sobering up, get a mug-up from the cook, poke around the woods talking to everybody (being a boomer he knew almost all of them) from P. F. to whistle punk. Then if the ground was too rough or he didn't like the donkey puncher or found the men didn't like the grub, he would roll his blankets and head for town.

At Loughborough Inlet, when Jack Phillips was pushing camp, 8-Day pulled the pin because a green whistle punk got his place at the table. Another time, a flunky handed him cold griddle cakes and 8-Day bunched it right now.

But at Cowichan Lake his line of approach back-fired. He had hired out to Matt Hemmingsen, completed his eight days but needing a good excuse to quit, he said he'd throw his boot up and if it stayed, so would he—for another eight days. The boot did—in a stove pipe hole—and Wilson stayed too. When he left he hung on a good one with a sixteen day stake.

CURLEY HUTTON

A pioneer logging railroader, Curley was driving an old 3-spot at Nimp-kish Lake, when it went through a bridge. He and the fireman managed to jump clear. "There must have been a hole in that bridge," English, the manager said. Hutton shook his head. "No hole then but there is now."

Curley also drove a lokey at Daddy Lamb's home guard camp at Menzies Bay—around 1921, was forever running over Lamb's prize pigs and being charged $35 each. In the days of the famous Goat Ranch at Alberni, there were two long trestles on the mainline—one low, one high. The bull bucker had been celebrating in port, and stopped at Goat Ranch to see the landlady, planning on catching the lokey back to camp. He got too much hospitality, missed his ride and bottle on hip, started to walk the ten miles.

On one of the trestles he fell between the ties and thinking he was on the 100-foot high one, hung by his fingers until he dropped—3 feet. He was so relieved he forgot the bottle which had jarred out of his pocket. Hutton and the fireman heard the story next day, found the bottle and a use for it.

Then there was the time Curley cleaned out the boiler of a Shay with three sacks of potatoes. The inside of the boiler was as clean as a dollar—the outside from cylinder heads to the last car decorated with foam like a Christmas cake.

BIG JACK MILLIGAN

Big Jack threw his hat on the skidroad and jumped on it with both feet. He'd hauled logs with bull teams. He'd roaded logs with donkeys. He'd rail-roaded logs with Shays. And if the bosses had given him elephants, he'd have skidded them out with elephants. "Hell," he said, "who ever heard of logging with trucks!"

This happened when Orville Milligan, Jack's brother and partner at Milligan's Camp on the Jordan River in 1919, bought a truck and fitted it out with bunks. But Jack was an old-school logger from the Quebec river drives. He was proud of his 11 x 15 Tacoma Roader known as "The Almighty Power" that dragged logs a mile and a quarter on a one-inch line. But rubber-tired trucks! He went on another two-week bender.

Bob Swanson worked with Big Jack when he was tending hook at Coburn's Camp, back of Nanaimo. "He was the biggest and most powerful man I ever knew," says Bob. "Two hundred and fifty pounds and six foot six in caulked boots. He was known all over the bush as the loudest hollerer on the coast. It was true. We were yarding on a 1,000-foot haul once and the two track-side settings were about a quarter of a mile apart on the same back-switch.

"One hot July day when the tree-boring beetles were wild on the wing, Big Jack was cooling off on a stump when one of those beetles bored into his foot, right through his boot. He let out a hoot so loud the whistle punks on both track-side machines shot in a signal that halted both yarders."

The trucks ruined Big Jack. After they started coming in and donkeys out, his stakes got shorter and when he got too old to tend hook (about 70) he took a job as a whistle punk. But he was still hollering his loudest just before he left for the Land of the Heavenly Timber in 1944.

NANAIMO NARROW GAUGE CLASSIC. The "Nanaimo," third locomotive in B. C. as she looked as a relic after hauling logs and lumber for Charlie Snowden's logger crew at Nanaimo in 1892. Engineer shown is J. K. Hickman. (Photo courtesy Robert E. Swanson)

BULL SLING BILL

If you shot straight with Bill Strausman you had a big-hearted friend in this colorful, all wool character—a legend in Northwest timber. A lady cook, new in a small camp, near Aberdeen in 1892, asked him what "P. F." stood for. "That's our cut, missus—practically finished."

Bill was logging in Michigan at 14—driving team on ice "wearing seven under shirts with the wild hay growing up the back of my neck." In 1910 he hired out to Saul Reamie at Rock Bay. Hanging his clothes on a nail, he found them on the floor in the morning. "A guy with whiskers to his knees said he'd used that nail for twenty years and if I wanted a nail to drive one in myself."

So Bull Sling Bill headed for the Charlottes. In 1915, he was camp foreman at Buckley Bay. Skagway Jack was pushing Camp 8 and Bill made the noise at Camp 9, on Masset Inlet. Bill reported they hired a crew of Haida Indians and the turns were hung up most of the time. In desperation Bill got down on his prayer bones and in a loud voice implored God to help—not to send his Son but come Himself as this was a man's job,—to come hell-for-breakfast in a dugout canoe and paddle this spear-headed crew clear off the claim. When Bill opened his eyes one Haida was halfway up a tree, yelling—"If God come down, I go meet Him!"

Then there was the fight with the Swede faller who wouldn't fall down. "Holy old Mackinaw!" Bill yelled. "Do I have to drive wedges in you and holler timber!" And the time a foreman told him crows could be taught to talk. "Talk? They can talk now—all of them. Listen there to them black devils— 'Yer broke! Yer broke'!"

In 1924 when Buckley Bay shut down, Bull Sling Bill hung on a big wingding in Vancouver and ran up so many bills he had to disappear. For twenty years while his debts were growing old, he grew old with them in a shack near Queen Charlotte City. Bob Swanson saw him at Cumshewa Inlet. "I'm in reverse gear," said Bill. "I go from foreman to hook tender to rigging slinger to choker man to whistle punk and now I'm a bull cook." He died in 1924 at Queen Charlotte City and was buried beside Box Car Pete.

HEWN TIMBER TRUCK ROAD showing Y-turn—Kelley Logging Co. at Church Creek, Cumshewa Inlet. (Photo British Columbia Provincial Archives)

SEATTLE RED

A South African who'd worked as a boy on a British Columbia ranch and then Abernethy—Loughheed's camp up the Fraser — how come he was called "Seattle Red"? Like this. He got tired of the Fraser camp and slipped over the border during a thunderstorm one night. He cached his pack and papers, start hiking north on the G.N. tracks. Stopped by Canadian Immigration, Red said he was from Seattle going to get a job in Canada. The officer disagreed. "You were but now, you're going right back to Seattle." Red did after retrieving his gear and went to work in a logging camp.

During 1926 Red claimed he worked in 35 camps in Oregon, Washington and California, staying only long enough to get a feed and wash his clothes. He had a passion for big automobiles and a philosophical sense of humor.

In one camp on the Island, he and Bob Swanson had to pack some Young Iron Works haulback blocks through three feet of snow. "If these here are young blocks I'd hate to pack 'em when they get full grown." That night Red asked the foreman if he had a horse collar. "If I've got to work like a horse I ought to be harnessed proper."

HARDY BULL SKINNER OF BIG WOODS. William Henry Barbrick who drove the skidroads of Vancouver and spent 60 years in B. C. timber. (Photo Courtesy Robert E. Swanson)

HARDY BULL SKINNER

Bill Barbrick drove bulls in the teak and mahogany of the Canal Zone and in the tropical heat of Central America. He knee-bolted in a Washington shingle mill in 1887, drove bulls in the fir—on the skidroads of Vancouver Townsite (the present Oak Street is one) and drank with other loggers at Gassy Jack's across False Creek.

When the bull teams died off, Bill worked on the booms and in 1912 was mate of a barge bringing logs from Queen Charlotte Islands. Before and after that he boomed for Bloedel, Stewart and Welch at Myrtle Point. Retired on a pension, he stood it as long as he could and then Sid Smith sent word to Menzies Bay to fix up a cabin for Bill. As Bob Swanson puts it—"He passed on to the camps of the Holy Ghost a few years ago—eighty-seven years old."

RED MORRISON

Old timers still talk about Red Morrison's Menagerie, how at 17, on his first job in the British Columbia woods as winter watchman for Dirty Face Jones camp at Elk Bay he held court to a wierd assortment of cats, wild Indian dogs, deer and cougars.

That was the way he broke in in 1902 but Red lived a full-rounded life logging. With a crew of ten he was sent north to handlog for Seaford's new mill at Judway—the first commercial logging venture in the Queen Charlottes. He was still going strong as high lead foreman on Vancouver Island when Bob Swanson worked for him in 1929.

SKY PILOT OF THE CAMPS

The British Columbia woods were not without The Word. Spiritual lines were passed out right along with chokers—in some places. There was the grand old saver of logger's souls—Reverend George C. F. Pringle—who spread the gospel in the woods for twenty-eight years.

Reverend Pringle headed toward the logger's heart in an ancient, leaky gas boat, "Sky Pilot" during the early '20s. He took over the Loggers Mission on the British Columbia coast. Experienced in the Klondike, such raw spirits as Rough House Pete and Canada Jack did not ruffle him. He knew most of the roughness was on the outside.

Like the day he struck into P. B. Anderson's Grassy Bay camp. His welcome was surly looks and guffaws. Then the bull cook, a fellow named Peasoup Brodeur, whispered in P. B.'s ear and Anderson immediately held out his hand to the preacher and asked, "Mr. Pringle, have you had breakfast yet?" It seemed the Reverend had saved Peasoup's life up on the Klondike.

On this show, Pringle saw what pliable stuff the logger's heart was really made of. He asked why a certain clump of big fir trees had been left in the clear. P. B. took him over and pointed up to a bird's nest. "There's eggs in it. We'll cut 'em after they get hatched."

BROOM HANDLE CHARLIE

That was what the kids called him—Bob Swanson was one—when he chased them off his brand new 27 ton saddle-tank engine back of Nanaimo. He was Charlie Snowden who ran the first lokey on Vancouver Island—the brass-trimmed 10 ton "Pioneer" built in England.

Charlie had worked on the coal trains from Departure Bay to the Wellington mines since he was 17. In 1906 he went into the woods as a donkey puncher but his first love won out when John Coburn built a sawmill at Cassidy and bought Charlie's old locomotive, the "Nanaimo." In 1908 a new sawmill was built at East Wellington and Charlie was picked to run an engine and the railroad.

One of the little engines got so haywire Charlie took it to the shop for a complete overhaul. After a leak developed in the water tank, the shop boss drained it, finding it stuffed full of shop tools. The foreman was burned at the idea of Charlie making off with the loot but undaunted, Charlie said to the bosses, "I'd have put the wheel lathe in there too only I couldn't lift it."

OLD HICKORY PALMER

From his early Idaho days through the roaring ones at Chemainus, E. J. Palmer was hard boiled but able and fair and gave many men like A. P. Allison, Matt Hemmingsen and others a sound start in life.

When he took over Chemainus, he laid steel on the skidroad and the Climax lokey dragged logs between the rails. In 1909 E. J. introduced the skidder system into the British Columbia woods. Before that it was Palmer's men who drove the famous Cowichan timber down river and who built a big flume from the lake to tide water. But Palmer was a railroader and had a line built in, which is being used today. By 1912 Edmond James Palmer's outfit was one of the biggest in the Province, cutting a quarter of a million feet in eight hours and became the sixth largest sawmill in the British Empire.

"NO FUN DOING NOTHING"

With that working philosophy P. B. (Peter Boward) Anderson rose and fell and rose during a busy lumbering lifetime. Broke in 1897, he towed twenty tamarack poles into Dawson to pay his debts. With 50c against a cheap meal at $1.50, he weathered it until he heard two men arguing in a saloon as to how to whipsaw lumber dumped off a boat. Pete Anderson had lost all his money in a sawmill deal in Bellingham but had learned a trade. "I'll set up the saw by morning."

He did and after the Klondike rush, landed in Seattle with a real stake. This was the spirit which guided and pushed the youth from Sweden to the Minnesota and Wisconsin woods where he worked for Weyerhaeuser and on to the Pacific Coast. Now he built a sawmill at Blanchard, Washington, and with 15 million feet of lumber on hand, the market went out from under him and he owed $34,000. He borrowed $250, built a mill out of round timbers in the Columbia River valley. His debts paid, he went to British Columbia. At Flat Island he worked on a contract to Hastings Mill Co. which was finished in 1916. Then with his sons, Dewey and Clay, he had his own outfit at Grassy Bay, employing many who later became well known superintendents and managers. (Writer Stewart Holbrook was timekeeper at this camp.)

The slump of 1929 broke P. B. again, but he had earned a reputation for integrity and the banks and wholesale houses financed the three men at Green Point Rapids. On his feet stronger than ever in 1937, the Anderson clan founded the Salmon River Logging Company.

MATT HEMMINGSEN

This was the man who in 1906 sent 15 million feet of logs rolling and leaping down the Tzolem River to the salt chuck. Humbird, boss of the Chemainus outfit forbid the building of a skidroad and ordered the river filled with logs. The logs got jammed tight and all attempts to loose them failed.

From Wisconsin came Matt Hemmingsen. He walked over the six mile log jam, sized up the situation and then blasted the rock bends out of the river. With the fall floods the jam broke and the last big blue butt went down to Courtenay, the booming ground six tiers deep, a million feet going clear out to sea.

Matt was also the man who rigged up the first skyline system ever used in the woods. This was in 1914 at Wardroper Bay on Cowichan Lake, the place he subdued or at least tolerated such rebels as Roughhouse Pete, 8-Day Wilson and Jesse James.

This was the same man who turned down $20,000 a year to run the Empire Lumber Co., who took a contract for E. J. Palmer and was stuck with 30 million feet of logs when the Chemainus mill burned down in 1923. Then after bad luck at mining and trying to develop a marble quarry, he founded the Hemmingsen-Cameron Timber Co. at Port Renfrew, taking over the old Sorenson outfit where alder grew up around the drums of 38 old donkeys and the railroad was crooked as a dog's hind leg. But that did pay and Mathias Hemmingsen retired to the green fields of Victoria.

STUB DILLON

Skid greaser and bull skinner in the Oregon woods at 18, Stub Dillon could do anything in the woods from setting a charge of powder to punching a donkey and that included fighting. He tangled with bull buckers, cooks and filers, in camps on the Cordova Street skidroad. The stronger and meaner they were the better Stub liked it.

Stub was a tough man in a tough world — hard worker, hard drinker, hard player. Out of his many scraps and scrapes came a real bruising — a fight over a woman in a bootleg joint in 1921. When his lady love sat on another fellow's knee, Stub landed on the victim. But a piece of pipe landed on Stub's head, shots were fired, the woman died and Stub got manslaughter.

WILDCAT OF THE WOODS

Most British Columbia loggers can readily name the man who saved his men the bother of quitting just to get a few drinks, by having a bar right in camp—the man who died eating a steak. Jesse James, of course—swashbuckling Jesse of James Logging Co., Keystone Logging Co. etc.

It is told around how one night in a Cordova Street saloon he hired Rough-house Pete Olson as hook tender, how they went to Victoria that same night, got a room and threw a party with a bathtub full of bottled beer. It is said Pete left the tap running while he got interested in one of the girls, the labels floated off the bottles, plugged the drain, a flood threatening the whole hotel.

Police were called, Jesse James and party put out on the street. Promptly hiring the girls as flunkies in camp, Jesse loaded one and all into his twelve-cylinder Marmon and headed north over the snaky Malahat at do-or-die speed. Day was breaking when they got to the foot of Cowichan Lake where Jesse's speed boat, "Stud Cat," waited. Twenty miles up the lake, camp was in sight when the "Stud Cat" hit a deadhead and went down. Pete and Jesse managed to save the girls because the camp sure needed flunkies.

But Jesse James who liked to live dangerously was soon to be cheated. Shortly after the Cowichan Lake fiasco, he was laughing at a joke in a Vancouver restaurant with his mouth full of rare beef steak and he choked to death.

WASHINGTON

Washington. You were back in the land of your fathers, silver dollars and gold teeth. You had seen the highballing wonders of British Columbia logging and now you were to see the domestic variety stepped up with American speed and organization. It wasn't as glamorous maybe but the logs and camps and mills were just as big and there were a lot more of them.

This was the green land which loggers from Maine to Louisiana had taken over, with Swedes and Finns from Minnesota, Michigan and Wisconsin included. This was home with the Big Woods flavor. This was Washington state where Jim Bridgeford had first tested out the merits of wire cable as a substitute for manila rope, where Capt. A. M. Simpson started the first sawmill on Grays Harbor.

You knew about the Pope and Talbots too—how they built a big mill at Port Gamble, others at Port Ludlow and Cosmopolis. Cyrus Walker was their first general manager and did things in the grand, stars-and-stripes way, building a big New England-type mansion on the hill, complete with American flag, picket fence, croquet court and brass cannon which fired a salute every time Walker's favorite schooner Forest Queen entered the harbor. They said at one time he had two hundred bulls in the woods and eighty-nine ships carrying lumber. And they said Pope and Talbot paid in cash every day or you left it in storage until you quit.

MALINOWSKI DAM. Historically famous in Grays Harbor County. Built in 1902 as Wishkah Boom Company's main river dam, it was rebuilt in 1917, stopped working in 1924. Operated for 15 years by Joe Malinowski, dam was used to create artificial freshets to float logs to tide water, about 20 miles. Gates were raised and lowered by force of water. (H. G. Nelson photo courtesy Rayonier Incorporated)

"WHOA, BETSY". Ten ton Porter rod engine, first locomotive acquired by Polson Logging Company and purportedly first to come west over mountains. In late years Northern Pacific Railroad, from whom it was purchased, traded Polson a large freight locomotive for it and placed Betsy in N. P. museum in Chicago. (H. G. Nelson photo courtesy Rayonier Incorporated)

GOOD HUNTING. Polson loggers save their skins for Sunday show. At left, holding dog, is Jake Snooks, next is Bob Blair, third is Frank McEachern —man in foreground unidentified. (H. G. Nelson photo courtesy Rayonier Incorporated)

BIG HOQUIAM AND BAGLEY SCRAPER— 1908. Photograph near Oxford's Prairie. Don-

That was a long time before you hit this country but the big shows of Simpson Logging Company guided by the Mark Reed clan were booming around Shelton. Weyerhaeuser, Long-Bell, St. Paul and Tacoma seemed to be logging half the earth, with Bloedel-Donovan, Merrill and Ring, the Polsons, Hama Hama and Schafer Brothers coming in strong. And there were a hundred and one smaller but still potent outfits around—like Siler, Cherry Valley, English, Dempsey, the Bordeaux brothers, Vance, Clemons.

You saw all this with the rattle of blocks, the crash of timber and the roar of mainline locomotives in your ears. You talked with photographers like Asahel Curtis, the two Kinseys, John Cress and others who recorded all this clamor and bang on plate. You talked with writers (and by that time you were selling stories about the woods and mills yourself) and began to realize daylight had come into the swamp for you and you liked it.

One of those writers was James Stevens, at that time known as the official recorder of the Paul Bunyan legends. Jim had been lumberjack, casual laborer, high wheel driver, sawmill worker—was later to become public relations head of the West Coast Lumbermen's Association, champion of the logger as a man and logging as a means of growing more timber.

Yet Jim was and is essentially a teller of tales. They have appeared in the Saturday Evening Post, dozens of other magazines, newspaper columns, and in several novels. The humorous sketch appearing on page 55 offers insight into the lumber worker's character and shows you Stevens' own sympathetic regard for the working stiff. It tells the story as a skilled writer saw one part of the "glory days of logging."

key made by Hoquiam Iron Works. Scraper was used to build railroad grade. (H. G. Nelson photo courtesy Rayonier Incorporated)

TRACK CYCLE OUTLASTS LOGGERS. John Lusk, time keeper at Polson's Railroad Camp, pedalled this vehicle like bicycle. It had friction brakes operating on axles. (H. G. Nelson photo courtesy Rayonier Incorporated)

FAMOUS OLD DOLBEER vertical spool donkey built by Murray Bros. in San Francisco, brought to Polson woods in 1887 by Cy Blackwell. (H. G. Nelson photo courtesy Rayonier Incorporated)

LOGGERS' PIG STY at Florence camp, near Index. Cook, housekeeper and pigs in front of south wall of logging camp pen on Snohomish River in days when logs were handier than lumber. A worse menace than wild loggers struck the Snohomish River pioneers about 1895. The Bohams had timber land near Falls City and wanted to farm. They cleared a little space to raise pigs but could not grow enough feed. Salmon were choking the river and they caught hundreds. But they quickly found the pork so badly tainted even the Indians would not eat it. The Bohams promptly gave up farming, turned the pigs loose in the woods and opened a store. The pigs went wild, rooted up all the farmers fields. It took all the white population and several seasons to clear the country of the bad pork. (Photo Darius Kinsey courtesy Mrs. Olive Quigley)

BIG GRUB STRIKE—1901. Bags, baggage and bedbugs, loggers quit Polson's table board for fatter feeding elsewhere. Some men traveled light, arriving in camp with two gallons of whiskey (75 cents a gallon) wrapped in a quilt, leaving with quilt only. (H. G. Nelson photo courtesy Rayonier Incorporated)

THE SHORT STAKE MAN

By James Stevens

The first day in camp is hell for the short stake man, the second is purgatory, and thereafter he gradually ascends to the portals of bliss, where a gray young worried timekeeper admits him to heaven with a timecheck. His prayerbook is the greasy thumbed timebook in which is scribbled the figures of stakes made on a hundred jobs, and his Sundays, evenings and holidays are spent in rapt porings over its pages.

We should not look at him on his first night in camp, when he sits with the gloom of a lost soul because he has only one shift to figure up. Rather let us gaze upon him when his anguish is subsiding and a bit of cheer is lifting the corners of his mouth.

His second supper in camp is done, and Short Stake comes into the bunkhouse with his first gay word. "Jist about got the wrinkles out now, boys." Hearing no outbursts of laughter, he moseys on to his bunk, digs into the cracker box nailed above his head, snakes out cob pipe, Onion Leader, snoose can, and, fondly, reverently, religiously almost, he brings forth the book of his soul, the paper of his heart, the shape and person of his ambition, the old wire-bound timebook. And the snag-headed pencil which is only sharpened once because it is chewed down to the point before the lead from the first sharpening is used out. He throws the stuff on his bunk; he fills his pipe and puffs; he stretches out, and for a moment he seems to drowse. But no. Soon his eyes open and he begins the dreamy blink. Unconsciously he blinks, unconsciously he puffs, for his eyes have a vacant, far-away stare; his consciousness is trailing over the hump to a far town, a good town, to boards white with jobs, to the train tha's going to the real camp, the good camp, the ideal camp where everybody is happy and satisfied and never wants to go anywhere else at all. There's grass around the bunkhouses there, and the cookhouse is cool, and the tables are slick and white. No brush in the woods, no rocks, no high-ball bosses, but there's —let's see—there's—

What was he thinking about? Short Stake comes back to life and rubs his chin reflectively. Oh, yes, he guesses he wants to shave. That's it. He sits up and begins to look for his razor. Now he'll have to borrow a strap and a brush and a glass, and maybe his soap is about played out, too. The timebook and pencil to catch his gaze. Well, the whiskers are not so bad. Let them go till tomorrow night. Wonder how much he'll have coming by then, anyway.

Off come his boots, then, carefully and with ceremony, the book is spread out and opened. Short Stake studies it glumly for a while, then the snag-headed pencil is lifted, the point strikes the paper and the first entry for this job is made:

8 hours at 45 cents a hour	$ 3.60
8 hours at 45 cents a hour	3.60

Then on another page:

Fare	$ 3.80
Hospittle	1.00
Snoose	.30
Onion Leader	.30
Sox	.30
Gloves	1.25
Shirt	1.50
Overalls	2.00
Caulks	.25
Thread and Needles	.25
Chocolate Bar	.10
	$11.05

Painfully Short Stake figures, and with caution and hope he goes over his sums again and again to be sure that he has not erred in the company's favor. But he has it right the first time: eleven dollars and five cents he owes, and seven twenty he has coming in wages. He sighs heavily; it is so hard to get ahead in this here world. Oh, well, says his optimism, he will be working for himself after tomorrow. But, no, he will still owe fifteen cents to the cursed company. It's a dod derned shame, that's all, for a man to have to sweat and grind and ache and starve and freeze for a few old rags and a railroad ride. What's the use of a man trying to do something for himself anyway when he has to suffer and sweat and grind and freeze—and that dod derned snoose can's about empty, and he'll have to get another one tomorrow night, which'll make him two bits in debt to the company. But—

But do not feel to sorry for Short Stake. He has his book and his figures and in them he finds hope and happiness. The pencil trots back and forth across another page and proves to him that by the end of the week he will have earned $21.60, and, if he is careful with his tobacco and snoose and his old sox, he will have at least nine dollars left after his fare, hospital and commissary are paid. At that rate he'll soon have plenty to get him to a real camp over the hump—the good camp, the ideal camp. . . . But hold! By the dying old hell-jumping gee whiz he's forgot board! Dod dern it, that spoils everything again. Here he was, happy and sitting pretty, feeling the warmest glow of content, and now he is out in the cold dark once more. Board at $1.15; more figuring; he'll be about even at the end of the week; and it will take him about twice as long to make that stake as he had figured it would. That will keep him here for . . . let's see, how much will he want to have when he leaves for the wonderful camp, the camp of his dream?

But whatever it is Short Stake will find consolation in the answer, for figures make him happy any time after his second supper in camp. He is the true bunkhouse optimist-dreamer. No camp has satisfied him yet, but he has no quarrel with "conditions." He has an ideal of a good outfit, and the dream leads him on. He is always headed for it, but he never finds it.

And he could be a much worse fellow. He seldom disturbs his bunkhouse mates with cranky diatribes, and he is no yeller of wild doctrines. Every night he spends pleasant hours over his timebook and with figures sustains his faith in the rare life he is sometime going to find. Magazines, books, pamphlets and newspapers may lie in piles around him; each of them may have many messages of inspiration from huge and wealthy organizations which throb and glitter with a passion to make him a noble man, an intelligent voter, a virtuous man in his pastimes, or a squawker, screecher, tom-tom beater, snake dancer and banner carrier in the war that will make every worker a boss in industry, even as the Mexican army makes every soldier a colonel; but they are all wasted on Short Stake, for he is too much concerned with his figures to read them.

The only way the uplift can reach Short Stake is to give him an adding machine and teach him to use it. This might improve the ritual of his hours by giving him increased figuring power; this would certainly add to his happiness.

A SPAR IS BORN

Equipped with special belt with life rope knotted in steel rings, long-spiked climbing spurs strapped to boots and legs, light axe and one-man cross-cut saw—the high rigger starts up. Life rope is manila line usually made with wire core, strong enough to hold the weight of a man under strain of jerking tree top yet capable of being severed with one axe blow if the tree splits and spreads, binding body to tree.

The high rigger works his way up with driving thrusts of legs, flipping the rope up eight or ten feet for the next hitch. There may be no branches on a spar tree bole for 70 or 80 feet from the base. Half an hour's work should see a good head rigger at the point where cut will be made, 150 to 180 feet up. There may be 40 to 60 feet or more of top, bush and crown.

First operation here is to make a deep notch at the point opposite falling direction of top. The second is to saw. When the tree shudders and top sways and starts its slow fall, the rigger drops his saw which hangs on a line, flips his life rope down a few feet, drops himself below the saw cut in case the top should kick back and sets his spurs deep, cinching the life rope tight. For full safety, spar topping is never done in a wind.

With deep ripping and splintering cracks, the big green crown leaps off the top and plunges down. The released spar pulls back, gathering speed, springs sharply, jerking and snapping back and forth in a 25 or 30 foot arc. Gradually the swaying diminishes until the amputated spar stands quiet.

The rigger may feel a certain dominance and conquest, may pause a few minutes, even stand on the flat top (3 or 4 feet in diameter), and clown a little. He has earned this show of eminence. But there is logging to be done and he goes down, letting the rope sag 10 feet below his spurs, swinging out and dropping, digging his spurs in and flipping the rope down again. A dozen hitches and he's on the ground. Another spar shaped up, ready for rigging.

When it is necessary to blow off a top, the high rigger saws and chops a girdle or shelf about 6 inches deep all around the tree. In this recess he packs seventy-five or eighty sticks of dynamite, about five sticks deep and sets a fuse six feet long. He may have rigged up a light block with line to come down on or he may come down by the usual rope-and-spur hitches. When fuse is lighted, rigger has five to six minutes to descend—"Plenty," as one high climber said with prime understatement, "if you don't get hung up somewhere." A well-set charge will blast the top off with almost as clean a break as a saw cut, barely bruising the spar more than six feet down.

HIGH RIGGER STARTS CLIMB. British Columbia lad looks confident and capable as he digs spikes into fir. (Photo Leonard Frank Collection, Vancouver, B. C.)

FIFTY FEET UP—NO LIMBS YET.　　　　**NOTCHING TO GUIDE TOP.** (Photos
Leonard Frank Collection, Vancouver, B. C.)

TREE SHUDDERS — TOP FALLS.   **TAKING FIVE BEFORE COMING DOWN.** (Photos
Leonard Frank Collection, Vancouver, B. C.)

DOWN WITH A WHISTLING WHUSH! (Photo Seattle Times)

"MAYBE I'LL JUMP — IT'S QUICKER". (Photo Leonard Frank Collection, Vancouver, B. C.)

THE HIGHCLIMBER

Highclimbing? Oh, there's not much to it;
    You could do it;
All you need is a bit of assurance,
    Lots of endurance,
And nerve enough for a try.
Look up there—up high,
Over the slope of that hill to the right.
    If you have good sight,
You can see a speck of humanity clinging,
    Swinging
In the bight of a rope, and trusting to luck
    And his pluck
And his spurs' firm grip,
To keep him from making a slip.
He's tired—and more,
His every muscle is sore
From the stress and the work and the strain,
    Till each moment is pain
And his legs grow numb and refuse
    Their full use;
Till his arms are all aches
And each blow that he takes
With the ax, or each stroke of the saw
    Rubs the raw purple flesh
To a fresh stir of agony, pent,
Where the spur straps are bent
    Round the knees.
He endeavors to ease the rack,
Leaning back on his belt to chop
The undercut in the top.

There it goes! See it leap
Gracefully down in a circling sweep!
    Look how he stands,
With hands all ready to steady
    'Gainst sliver or split,
    When undercuts hit!
The fling of the top sends him reeling,
Robs him of sense and of feeling,
And hangs him inert for a breath or so,
With death two hundred feet below.

See! He's descending, glad at the ending;
Look at him recklessly leaping below
    To the length
Of his rope, and tempting its strength.
Look at him, there,
    Beside the tree,
As straight as it, as sturdy, as free!
Deep in his heart a melody stirs—
A nameless melody, bold and fierce;
Bold as his courage and will,
Fierce as his toil and its thrill;
Voice of the savage lust of strife,
Chanting his wild young love of life;
    Jubilant, glorious,
Paean of youth victorious!
See him smile as he strides away!
He gambled with death and won—today.

CHARLES O. OLSON.

4 L Bulletin, May, 1922

THE SPAR AT WORK.

GOLD RUSH STARTED OLYMPIC AREA LOGGING

"When the gold rush hit California in 1849 and mining camps sprang up overnight," said the Port Angeles News of November 28, 1953, "lumber was needed for building the boom towns. There were no harbors tapping the Northern California timber, so California turned to Washington for fir.

"Ships from the east coast converged on San Francisco, and first sought hewn timbers to be carried to California for sawing into lumber. By the time Washington had become a territory, early settlers already were doing a thriving business in sawing lumber for shipment from Puget Sound.

"Great Douglas fir timber stands marched down to water's edge on Hood Canal, Puget Sound and the Strait of Juan de Fuca, and all their inlets. The Sound was a great harbor and its coves and inlets provided safe anchorage for sailing ships. Settlers along the Strait and Sound cut and squared the timbers, schooner skippers bargained for them at water's edge.

"Homesteaders skidded the logs to salt water with their ox teams, boomed them and waited for the ships. Much of the trade was by barter, as cash was scarce. Payment was mostly in trade goods brought north aboard the schooners and square riggers. The goods received kept the homesteaders going until they completed their homes. The logging operations cleared the land so it could get into production for farming."

In 1852 the first sawmill was built at Port Ludlow, another the next year at Port Gamble, and subsequently mills at Discovery Bay, Port Townsend and Port Blakely. To keep the big mills operating, logging graduated from the realm of small operations by individuals getting out a few logs for trade and land clearing.

"To insure a log supply, the mills established their own logging camps or contracted with logging operators. Many of the mills were built in large timber stands, with skid roads leading directly from the woods to the booming grounds. First logging was confined to shallow areas on level benches near tidewater or rivers. This was because ox teams dragged the logs from the woods over skid roads, making long haul operations impossible.

"The logs were dragged by oxen over skid roads, either to the Dungeness River and floated into the harbor, or directly to the bluffs, from where they were rolled into the water. Old-timers in the Dungeness valley imported and raised oxen for use in the woods. They are credited with raising a strain of cattle much sought after as oxen.

"Much of the fir in the valley and on the easily accessible plateaus above the valley and on the bluffs was logged and sent to Port Discovery and other nearby mills. About every early settler in the east end of Clallam County either did logging or worked in the Port Discovery mill in those days.

"As the valleys and plateaus in the east end of the county were logged off, logging advanced westward along the Strait of Juan de Fuca, although still confined to easy 'shows' where oxen could be used in the woods. The age of railroad logging had not yet arrived.

"First record of logging at Port Angeles was in the 60's, when Alfred Lee and his brother-in-law, C. W. Thompson, logged at Lee's creek and what is now Gale's Addition. Settlers may have done some small scale logging before that time.

"Lee and Thompson used oxen in the woods, and dumped the logs in Port Angeles Harbor, from where they were towed to up-sound mills. Thompson, a Dungeness rancher, continued his logging operations and is probably the first man to log west of Port Angeles.

"In 1883 Thompson started logging at Port Crescent. Remains of his old skid road still remain in the woods adjacent to the bay. The logs were boomed in the bay and towed to the mills.

"Next logging operation west of Port Angeles was that of the Hall and Bishop firm at Gettysburg, a logging settlement east of the Lyre River. That firm operated for a number of years on the fine stand of timber west of what is now Joyce. One member of the firm was the late State Senator William Bishop of Chimacum.

"While Hall and Bishop were operating in the Gettysburg area the steam logging donkey came into use, partially replacing oxen and horses in the woods. The donkeys were fired by wood and had steel cables on a drum, which pulled the logs along the skid road.

"Shingle production started in 1887 when the Puget Sound Co-operative Colony built its mill at Ennis Creek on Port Angeles harbor. Later that year the first Dry Creek Mill was started, the second in this area in 1889 by the Port Angeles Shingle and Lumber Co.

"BIG MILL"—PORT ANGELES. Built in 1914 by Michael Earles, mill was operated as Puget Sound Mills and Timber Company, later purchased by Charles Nelson Company who operated it until 1929. (Photo courtesy Port Angeles News)

"Logs at first came to the colony mill from land adjacent to the government townsite and from the Frank Chambers clearing on his homestead near the present golf course.

"In 1890 logs for the colony mill were coming from near the Lyre River, where they were cut at a camp operated by C. F. Clapp and C. W. Thompson. The mill closed down early in the 90's. From 1888 to 1890 Allen Meagher and Hartley Goodwin logged in the Dry Creek district. They operated with oxen and dumped their logs over the bluff into the lagoon at the head of Port Angeles Harbor. Among the tracts logged was the H. M. Guptill homestead located at Dry Creek, in 1886.

"Michael Earles, who later was to be a prime mover in Clallam County logging and sawmill development, first started logging and operating a sawmill at Clallam Bay in 1890. The mill operated two years, then closed, and later burned down.

"Logging in the Port Angeles area got its biggest boost when the Filion Brothers, Michigan mill men, built their mill on Lincoln Heights in 1893. Filion loggers started cutting the heavy stands of timber west and south of the city. Soon a logging railroad was enlarging the cutting area that extended far up into the lower foothills south of town, in the Black Diamond district, Little River valley and the big area south of Dry Creek and east of the Elwha river. The Filion logging railroad was the first of its kind near Port Angeles.

"The mill cut lumber and shingles, and was the city's only large industry for many years. Some horses were used in the woods, but steam donkey engines were in general use by then. The horses and donkey engines yarded the logs over skid roads, then they were reloaded on cars and hauled to the mill by steam locomotives. The mill operated for almost a half century under the Filion family and J. E. Beam ownership.

SHINGLES TO MARKET—1893. Three-horse team hauling shingles from a Dry Creek shingle mill near Port Angeles. (Photo courtesy Port Angeles News)

FIRST CO-OPERATIVE MILL. P Sound Co-operative Colony mill at E Creek built in 1888 near Port Ange (Photo courtesy Port Angeles News)

BOOMING GROUNDS, CRESCENT LOG-GING COMPANY at Francis Street, Port Angeles. (Photo courtesy Port Angeles News)

PORT CRESCENT IN 1902. Log dump of Puget Sound Sawmill and Shingle Co., Crescent Bay, near Port Angeles. (Photo courtesy Port Angeles News)

"In 1896 another large logging operation came to Clallam County when the Puget Sound Sawmill Co., headed by Michael Earles, started logging on a big scale at Port Crescent.

"The Port Crescent operations were the largest in the county. Earles revived the town of Port Crescent and made it into a boom community again. By that time steam donkeys had come into general use, but logs were still hauled along skid roads. The operations crept away from the salt water up toward the foothills on the big plateau south of Port Crescent and west of the present community of Joyce.

FIRST SHINGLE MILL IN CLALLAM COUNTY built by L. T. Hayner and associates—son Ray, William Graham, Clarence Mc-Loughlin, Nicholas Meagher, Jr., and Frank Patton—at Dry Creek, 1887. Shingles were hauled to steamer Evangel at Port Angeles and shipped up Sound. (Photo Port Angeles News)

"Logs were either cold-decked or hauled to reloads, and trains hauled the logs to the Port Crescent booming grounds. Until 1912 the logs were towed by company tugs to Bellingham sawmills. In 1914 Earles built a sawmill in Port Angeles, known as the 'big mill', toward the west end of the harbor. When the Milwaukee Railroad built west out of Port Angeles in 1914, Earles logged along the line between here and Twin, and started shipping his logs here on the railroad. Earles later sold the mill to the Charles Nelson Co., which operated it until 1930.

"At the turn of the century the Peterson family started logging in the Dungeness-Carlborg district of the east end of the county, dumping logs into Dungeness harbor. They operated the first logging railroad in that area. C. J. Erickson, the contractor for construction of the Milwaukee Railroad, bought out the Peterson family interests and built the Carlsborg mill in 1916. A logging railroad out of Carlsborg tapped the big timber stands south of Sequim and Carlsborg far up into the foothills. Lumber from the mill went over the Milwaukee road. Erickson operated the mill until 1934.

"About this same time the Snow Creek Logging Co. started activities. The company operated south of the head of Discovery Bay and dumped its

HIGH BALL CREW AT PYSHT in "glory days" of logging, about 1930. Merrill and Ring men include Ernest R. Cogburn, center, hands in pockets, 26 years old. Born into a logging family, Cogburn started tending hook for M & R at 19, went on to become a logging contractor in Alaska, then partner with Elmer L. Critchfield. (Darius Kinsey photo courtesy James Kerr)

EVERYBODY RIDES. Going back to camp is this Hall and Bishop crew at Gettysburg, Clallam County. Extreme left on engine — Jack Sands, fourth from left, George Sands, extreme right, John Udd, others unidentified. (Darius Kinsey photo courtesy J. L. Sands)

logs into the bay. The company logging railroad operated along the Jefferson and Clallam County boundary, with the camp in this county. The Snow Creek Co. cut many million feet of logs south of the head of Discovery Bay and Blyn. Operations ceased in the 1930's.

"Merrill & Ring started logging at Pysht in 1916, continuing until 1944. During that time the company logged about 25,000 acres south, east and west of Pysht. The company used steam donkeys and a logging railroad, and dumped logs into the mouth of the Pysht river. The company hauled log rafts to up-sound mills with its own fleet of tugs, the Wanderer, Walloha and Humacomma.

"The Bloedel-Donovan Co. started logging out of Sekiu in 1924. First logging was in the area south of Clallam Bay. Soon the company built a logging railroad over the hill through the west Pysht Valley, to tap the timber along the Soleduck, Calawah and Bogachiel Rivers.

"This company was really getting up into the high hills, and the high rigging or sky line logging became general. The spar trees were put on the hill tops, and logs from the side hills and valleys were hauled up, loaded on trucks are taken down hill to the railroad reloads.

"The company operated many "sides", with camps at Sappho, Beaver and Hoko, and headquarters at Sekiu where the railroad shops were maintained. By 1941 the company was at its peak with 1,000 loggers on the payroll. Logs were towed to the company mill at Bellingham.

"Bloedel-Donovan sold out to Rayonier Incorporated in 1945 and the company logging division has operated it since. The present operations and

HIGHEST PILE TRESTLE in the world said Darius Kinsey who took the picture near Hamilton on Lyman Timber Company operation in 1918—136 feet high. (Darius Kinsey photo from Jesse E. Ebert Collection)

the late operations of Bloedel-Donovan were getting further and further away from salt water, and out of private timber into the Olympic National Forest.

"In 1917 the U. S. Spruce Division started searching for airplane spruce in the west end of Clallam County, and built the Spruce Railroad to Lake Pleasant. The railroad joined the Milwaukee Railroad at Disque, west of Joyce. The costly railroad hauled little spruce, as the Armistice came before a mill at Ennis Creek was completed.

"A new figure entered the Clallam County logging picture when the late Petrus Pearson came here as secretary-manager of the Irving-Hartley Logging Company. The company started logging in the Twin Rivers area in 1924. A logging railroad was built to feed logs to the Milwaukee Railroad, which then terminated at Deep Creek. The company logged the bench below the higher foothills, between East Twin River and Deep Creek. Logs came here and to the Carlsborg mill. Some were boomed here and towed up-sound.

"In 1927 the company was reorganized as the Crescent Logging Co., with Pearson as vice-president and manager. The Piedmont (Lake Crescent) camp was opened. Logging was in the area west of the lake and south of Joyce. Next operation was at Riverside, in the lower Calawah and Soleduck

SPIDER-LEG BRIDGE but strong enough to transport train loads of logs for Maine Logging Company. (Darius Kinsey photo from Jesse E. Ebert Collection)

river areas. These two camps fed logs to the Port Angeles Western Railroad, which the company had acquired from the Lyon-Hill Co.

"The Crescent company extended the railroad 7.4 miles from Lake Pleasant almost to Forks. Tractors and logging arches were used in the woods by the Crescent Logging Co. as early as 1930. Its last operation was at Camp One, immediately west of Lake Crescent in the upper Soleduck valley and the hills west of the valley. The company holdings were sold to Fibreboard Products Inc. in 1946.

"Pearson and associates bought the Carlsborg mill in 1940, and timber from Crescent Logging Co. holdings went to that mill and a shingle mill at Lake Pleasant.

"Neah Bay logging operations started in 1926, when contractors started cutting, mostly spruce on the Makah reservation. The logs were cut and split into cordwood length, and went to the bayshore by a small logging railroad. The wood was cut up in a chipping plant and shipped to the Crown Zellerbach newsprint mill here. In 1932 Crown Zellerbach acquired timber on the Makah reservation and started the Sail River camp. A logging railroad was built and the logs hauled to Sail River where they were boomed and towed here."

TRESTLE AT ROBINSON'S CAMP—Clallam County, Washington. (Photo Axtell courtesy J. L. Sands)

MILWAUKEE "Y" AT TWIN Puget Sound Mill and Timber camp at end of Milwaukee Railroad about 1918. Family houses at left and right, logging camp center. (Darius Kinsey photo courtesy Critchfield Logging Company, Inc.)

REMEMBER FLYING THOMPSON?

There were sixteen legendary characters in logging's heyday for every straight-down-the-middle boss who saved his money, used his brains and became a leader in the industry. And this sixteen, multiplied for sake of argument by four hundred and ten, furnishes plenty of material to keep old-time talk going around the hot stove for a long time yet.

For instance there was Step-And-A-Half Jack Phelps. When he fired a man he'd ask if he had a partner, and if so he'd fire both of them so they wouldn't get lonely. Moonlight Joe of Idaho would howl like a banshee for "Logs! More logs!" Boomer Todd of Saddle Mountain once found a bear in his kitchen, chased in by prankish loggers. He killed the bear with a cleaver and the boys ate bear meat that night—all except those who put the bear in. So to keep the legends lively, forget about the men who built empires and hark to those stalwarts who thought empires were in baseball and couldn't spell the word anyhow.

Elmer Critchfield and Ernie Cogburn remember Jimmie, The Fish. His real name was Jim Salmon and he was a rigging slinger on the Olympic Peninsula. Dirty Shirt Jim they also called him because he was never known to take a bath or change his shirt. The last Elmer and Ernie heard of him was when the other loggers forced him under a shower and gave him a clean shirt. Jimmie headed out that night.

And the millionaire hooktender—Frank Lowry? He owned six or seven suits and wore spats on his town tours between jobs. He was killed near North Bend when the rigging hit a snag. There was a timber cruiser with a similar name—Frank Crowley who really was a millionaire—almost. He had about $150,000 in safe deposit boxes and spent his later life not spending the money.

Ever hear of Old Ninety? C. L. Bowie always said he was exactly ninety years old. He was an old man with scars all over his neck when he

ALMOST STRAIGHT UP. Note the detail, like a Japanese curtain, in this photograph of 66½ percent incline grade— Wisconsin Logging and Timber Co., Columbia River, Washington. (Photo Oregon Collection, University of Oregon)

3300 FOOT INCLINE up to 40% grade on McNeill-O'Hearn and Company operation. (Darius Kinsey photo from Jesse E. Ebert Collection)

went to work in the Simpson camps as a skid saddler and never seemed to age, they said. He kept a loaded gun in his bunk but liked to dance and would walk three miles to show the young punks how to cut a rug.

And Celestial Sam, the mild and apologetic little bucker with a long, sharp nose and flat, sallow face. He was that solemn character told about many times who carried the Ingersol watch, carefully protected, in a snoose box with hole punched in it for the stem. Loggers would ask—"Well, Sam, what time you got by the snoose box?" And Sam would plunge his hand deep in his pocket, dredge up the box, remove the cover with a last-act reverence and gloomily intone the hour. Then always he would add — "And so much nearer eternity!"

From somewhere back in the dim and distant comes the memory of the bull cook who played a poker system with a blind and steadfast faith but was always broke three days after the eagle passed over and had to stay in camp when the others went out. He would say sadly—"Some day I'll hit it and go out with you fellows." Nobody remembers if he ever did.

MAIN LINE ON GROUND LEAD. Road donkey seen at crest of hill. Haul back line at right of road, stump roller at left. (Darius Kinsey photo courtesy University of Washington)

LOADING SCHOONER AT TACOMA. It was hand work in 1916 getting lumber aboard and below. (Photo Oregon Collection, University of Oregon)

Stewart Holbrook tells about Frank, the B. C. logger who was once run out of Big Bill's saloon for his arrogant, insulting talk. About a month later Frank came back, friendly and peaceful. Big Bill served him a drink. Frank leaned over to the logger next to him at the bar and reached inside his shirt. "You carry that around with you all the time?" He had pulled out a live snake which fell writhing to the floor. Big Bill was dumbfounded but almost ready to give Frank the heave-ho once more.

Then Frank plucked a rat from around the end of the bar. Big Bill started for him and Frank made a pass at Big Bill's thick mat of hair. A baby octopus magically appeared in his hands. "Good. Fine for soup," said logger Frank. "I keep him." Big Bill snapped out of his trance and made a grab for Frank, roaring—"You trying to make a monkey out of me!" Frank stepped nimbly aside. "Not me. No. Your pa and ma. They did it."

Everybody in the saloon had backed up expecting fireworks and to see Frank separated into little pieces. But instead, there suddenly appeared from the logger's mouth a stream of fire and smoke—a beautiful blue and orange flame. Big Bill stood transfixed as Frank tipped his cap. As he left he said quietly: "You see before I get into trouble I am The Great Precelle, unexcelled prestigidator of all Europe."

FIRING DONKEY WITH PEELER LOGS at Goodyear camp in Clallam County, Washington, just after being taken over by Bloedel-Donovan. Wood cutter bucking prime fir, wood splitter making slabs. Donkey puncher at right is Elmer Critchfield, lifetime Washington logger. 2 million feet in cold deck. (Photo Darius Kinsey courtesy Critchfield Logging Company, Inc.)

NORTH BEND SYSTEM across Skagit River — Faber Logging Company. (Darius Kinsey photo from Jesse E. Ebert Collection)

THE RIGGER

With steel-caulked shoes
That bite and grip,
With their laces hid
So they cannot trip;
With slicker short, for the sake of ease,
And tin-pants stagged, close to the knees,
He takes the trail among the trees.

The rain comes down
With a sidelong sweep;
The branches softly sway and weep,
And make a pool
Where the trail is deep.
He casts a vigilant glance on high—
Overhead—where the fir-tops sigh;
The water pours from his glazed hat brim
As he looks aloft from under the rim.

He leaps a puddle and swears a bit,
At the wet and the wind
And the chill of it;
Rivulets trickle and flow and leap,
And cascade down
Where the ground is steep;
He wades through mud where the yarder stands,
And climbs to the boiler
To warm his hands;
He grabs his gloves, takes a chew of snoose,
And joins the crew
That by threes and twos
Plod out to their work through the muck and ooze.

TRAM ROAD CROSSED GULLEY after meandering down hillside at Coal Creek Logging Company operation near Maple Falls. Cars were snubbed down grade. Note overturned car in brush. (Darius Kinsey photo from Jesse E. Ebert Collection)

The yarder-drums stir,
And the cables glide
Up through the high-lead and down beside;
Liquid, enveloping, yellow-brown mud,
Clings to them, covers them,
Drains in a flood
From their thread-like lengths,
As they tighten and slack
Over the road to the woods and back;
From spar-pole blocks
Comes a steady rain
Of spattering slush, as they swing and strain.

The loaders stand knee-deep
In debris and mire,
Over a sodden, smoke-blanketed fire
That sulks and sputters,
And will not burn.

The chaser sits, hunched, awaiting the turn;
All of his garments are slushy,
And down
From his paraffine pants
Stream small rivers of brown;
Where the sticky clay stain
From choker and chain
Dissolves in the wetness and colors the rain.

Incessant—depressing—benumbing—and chill,
Drizzle and splash,
Over yard, over fill—
Above and beyond—to the end of things,
Where the vision ends
And the cloud bank clings.

—CHARLES O. OLSON.
4 L Bulletin, March, 1922

LOADING AT HALL AND BISHOP'S—GETTYSBURG, 25 miles west of Port Angeles. Hall at right in front of 40 ton Climax. (Photo Axtell courtesy J. L. Sands)

LOGGING COTTONWOOD within present city limits of Sedro-Woolley—first logging venture of father of Ernest Cogburn after arriving from North Carolina. (Photo Darius Kinsey courtesy E. R. Cogburn)

MERRILL AND RING CREW AT PYSHT. Joe "Bill" Vane, Mark Lauchran, Lawrence Higginbotham, Tom Newton and far right, Jim Higginbotham. Newton was high-climber. One noon hour he was climbing over limbs 125 feet up without life rope, missed his hand hold and fell to the ground. He merely picked himself up, walked into the cook house and had lunch. He was killed by crane at the building of Coulee Dam. (Photo Darius Kinsey courtesy Critchfield Logging Company Inc.)

NORTH BEND WOODS CREW and Willamette yarder. Lyle Stow, superintendent, extreme left, and crew of North Bend Timber Company in 1937. (Photo Darius Kinsey courtesy Lyle Stow)

2-LOG TRAIL CHUTE with corduroy road for trail team—St. Joe National Forest about 1911. Point in foreground blasted out for chute. (Photo courtesy U. S. Forest Service)

IDAHO and MONTANA

LUMBERING IN THE ST. JOE FOREST

About 1898, a rush for timber claims took place on the western and more accessible portions of the St. Joe Forest in the "Panhandle" of Northern Idaho. The first sawmill came in 1889, a small one compared to those that were built at Harrison and elsewhere on Coeur d'Alene Lake, but it marked the beginning of a 60-year period of logging and milling that saw, at its peak, six large mills operating within or adjacent to the forest. Much of the heavy cut of timber was made possible by the connection of the St. Maries River section of the Chicago, Milwaukee, St. Paul and Pacific Railway with the main line in 1909. However, river driving and rafting accounted for the removal of much of the early day volume, especially from Marble Creek. The Elk River branch of the C., M., St. P. & P. Railway, following the course of the St. Maries River, moved more sawlog volume than any branch line in the Inland Empire.

The first large sawmill was installed at Potlatch, Idaho, in 1906 by the Potlatch Lumber Company. Following its installation came one at St. Joe City in 1908, primarily to supply construction timber for the railroad. A second Potlatch Lbr. Co. mill was installed at Elk River in 1910. Two mills were built at St. Maries, one in 1911 and one in 1912. A mill was built at Fernwood in 1913. During the heydays, from 1920 to 1930, frequently over 300 million board feet, mostly white pine, was harvested annually. The largest proportion of this cut came from privately owned lands.

UNLOADING AT LANDING—KANISKU NATIONAL FOREST at Beardmore timber sale, 1923. (Photo K. D. Swan courtesy U. S. Forest Service)

LANDING AT CALIFORNIA CREEK Sale Area.
(Photo K. D. Swan, courtesy U. S. Forest Service)

(Opposite) BIG CREEK FLUME emptying into Priest River, 1927.
(Photo courtesy Ennis Day and Charles H. Scribner)

PRIEST RIVER LOG DRIVE

"The West's only major log drive is drifting toward an end which is scheduled for mid-June," said the Christian Science Monitor of June 18, 1948. "Nimble footwork is needed as 'pig's' ride herd over 23 million feet of logs.

"Ordinarily, there are two traditional drives remaining in the old West: one on the Clearwater River, and the other on the Priest and Pend Oreille Rivers of northern Idaho. But wet weather reduced forest production this spring and the Diamond Match Company drive of 23,500,000 board feet is the lone reenactment of this ancient custom of floating logs to the mill.

"For 16 years Bill Whetsler has been foreman of the annual Diamond drive from the white pine forests surrounding Priest Lake, Idaho, down the winding Priest and Pend Oreille Rivers to the mill 67 miles away at Newport, Wash. 'This year's drive is my second biggest,' Mr. Whetsler observes. 'Last year, we took out 28,000,000 feet.'

"Driving gives a thrill to the men who ride the logs. Most of the crew of 18 have been at it since youth, two for more than 50 years each: Paul La-Motte and Henry Lund who drove logs in Norway before coming to the United States. Foreman Whetsler has been riding the booms since he was 15. 'In Norway as a boy,' says Mr. Lund, 'I ran really rough rivers. And in 1913, when I first came to this country, these drives were harder. We got $3 a day then for 10 or 12 hours' work.' Now the drive crew receive $1.75 an hour for 40 hours plus overtime.

LOADING CARS FORE-AND-AFT LANDING—Spring Gulch, Lolo National Forest, by Montana Logging Company, 1923. (Photo courtesy U. S. Forest Service)

"The Priest River log drive is spectacular in spots. Since the first drive in 1901, when cedar poles were transported three miles down the river, many of the kinks and rapids have been eliminated from the drive's course. Today, four major rapids remain: Chipmunk, Binarch, Eight Mile, and Finstad. 'After those, we just float home easily,' says Mr. Whetsler. The tiny logging town of Priest River, at the confluence of the Priest and Pend Oreille Rivers, schedules a celebration near the end of May about the time the drive passes the town.

"The logs are felled during the winter and spring, and propelled into Priest Lake on long forest slides. They are rounded up by boat and chuted into Priest River, where the current carries them toward the mill. But in shallows and rapids, and occasionally for no apparent reason, the logs jam with as many as 11,000,000 board feet clogging the swift river.

"Sometimes the jam lies next to the shore; this is called a 'wing.' Sometimes the jam builds in the center of the river, an island of logs with rushing water on all sides; then it is a 'center.' Both must be broken up by the 'river pigs' carrying peavies, the traditional metal-tipped staffs of the logger.

"The men use the peavies to pry loose logs at the very front of the jams until they find the nest holding the entire pileup. Then the work becomes dangerous and, in a rapids, spectacular, because as the men loosen these key logs, the big ones behind begin tumbling down the river. The 'river pigs' scramble and leap along the rolling logs to solid shore or a nearby jam that has not budged. To reach centers, the men row a bateau, a flat-bottomed river boat. The technique is identical; they loosen the key logs and then hurl themselves into the bateau as the logs begin to churn downstream."

MARTIN OLSON DRIVE IN FISHHOOK CREEK—St. Joe National Forest.
(Photo courtesy Charles H. Scribner)

LOG JAM—MARTIN OLSON DRIVE in Fishhook Creek—St. Joe National Forest.
(Photo courtesy Charles H. Scribner)

ROLLING LOGS INTO FLUME—Skookum Creek sale, Coeur d'Alene National
Forest in 1924. (Photo K. D. Swan, courtesy U. S. Forest Service)

LUMBERJACK FIGHTS

by John C. Frohlicher

"Kalispell hop!" Ringing clear through the heavy, smoke-filled air of the Butte speakeasy, I heard the battle cry of the "Timber Willie" from my own northwestern Montana logging country. I turned from the bar, and saw him, the lumberjack, a strangely garbed figure in that room of pale-faced, over-all-clad miners. I knew the significance of that shout. Turning to the bar, I said, "Straight, Jack."

The rest of the crowd, mostly newspaper men who had dropped in for a nightcap, looked askance at the logger. He snarled, "Don't you scissorbills know enough to belly up when you see the 'Hop'?" Then he shouldered his way in beside me and said, "You may have four eyes, and live in a haywire camp like Butte, but you sure know loggers. Make mine the same, barboy." He bought and I bought. A typical logger, I thought; his pockets full of hot money and his eyes full of devilment.

My friend was pretty well oiled when he came in. So after the second drink, he decided to see if the bar was real mahogany. Up on top of the mirror-like bar he leaped, and, digging in his caulks, started to execute the Kalispell hop. Two big ex-miners of Butte materialized out of the back room and tried to take that timber beast apart. (I say "tried" advisedly.)

It was a beautiful mill. The miners were willing to fight, but their bellies were full in a hurry. The barboy called the police, but before the law arrived the bouncers were insensible on the floor.

Another reporter and I hustled my wildcat friend out a side door and into a taxi. We reached the station a minute before the westbound train pulled out, and put him in the smoker. He went peaceably; the fight was apparently out of his system. But he called to me as the train left, "Them Butte buzzards ain't so tough, kid!"

That was my latest sight of a logger in physical combat. As I rode back up town I grew reminiscent—thinking of the old days when I was proud to be called a "timber beast".

I saw my first lumberjack fight when I was about six years old, and the animal logging was in its heyday around Kalispell, Mont. The John B. O'Brien Lumber Co. was operating at Somers, on Flathead lake; the Northwestern Lumber Co. was sawing at Kalispell, and a flock of portable outfits were butchering timber here and there around the valley.

BENEWAH CREEK FLUME—1915. In St. Joe National Forest.
(Photo William Roddy, courtesy Charles H. Scribner)

UNLOADING 8-WHEEL TRUCKS—in Kootenai National Forest by Neil Lumber Co., Libby, Montana, 1921. (Photo K. D. Swan, courtesy U. S. Forest Service)

18,000 FEET OF LARCH NEAR KALISPELL at Henry Good Camp in 1928. (Photo courtesy U. S. Forest Service)

MAIN LINE AND LATERAL. Donkey trailing logs, with power for both main line and spur chutes. (Photo courtesy U. S. Forest Service)

One Saturday afternoon, Bill Brandner, foreman of an O'Brien camp, engaged in an argument with George MacCauley, push for a Northwestern layout, in one of the score of saloons that lined Main street of Kalispell. Evenly matched men they were, six footers, 200 pounds in the buff, both crack rivermen and cats for quickness. The row started in the Silver Dollar saloon, but the push of eager spectators shoved the battlers out in the street.

They grappled in the mud in front of the saloon, their faces ruddy and distorted with hate. George depended on blows and kicks, while Bill used wrestler's tactics. A kick in the groin doubled Bill up for a moment; then a heavy blow on the jaw knocked him over. But his enemy was a little slow in putting the boots to him, so Bill got up, his face unscathed by the caulks, and fastened his hands on George at last. A hammerlock. Up—up—the arm was forced. "Are you licked?" gasped Bill.

"Hell, no!" cried George. The arm snapped—a dry, brittle sound, like the crack of an old pine-cone broken under a heavy boot. George fainted.

A cop stepped up and arrested Bill. Bill protested. Presently George came to and cried, "Help me up, Bill, an' we'll lick that thus-and-such right!" The arrest didn't hold. None but an ex-river pig could arrest a white water boy fresh off the jam and get away with it. And the two ex-pigs on the force had sense enough to stay at the station and play cribbage when the drive hit town.

During my boyhood I saw the "Line" in Kalispell taken apart several times when the drive hit the dam. Then they drove the Flathead, the Stillwater and the Whitefish rivers, all joining within three miles of home, and drove the Swan river, 30 miles away. Bigfork, where the Swan flows into Flathead Lake, was as hard a little town in its day as any of the old Wisconsin towns, so my father, who saw the Menominee when 50 different drives went through the sorting gaps there, tells me. It was at Bigfork that I witnessed one of the finest spectacles of my life, when a man who weighed about 140 pounds whipped a woods-bully who must have weighed 190.

Charley Jungberg, the little man, was at that time walking boss for the O'Brien outfit. He was being vocally abused by the bully. Both men were cold sober; indeed, I know that Charley never drank. But the big fellow thought he had something easy to pick on. Charley was sitting on the porch of the hotel. The bully grew more violent in his remarks and finally aimed a swing at Charley's head. The little man exploded.

His first offensive was launched at the shin of his challenger. Charley wore the caulked boots of the riverman, and the caulks bit deep. The big man gasped, but came forward, and his fist connected with Charley's jaw. The little Swede recovered in an instant, and rushed. He ducked two swings, and made the prettiest tackle I've ever seen off a football field. Crash! His husky opponent lit like a thousand of brick, and Charley was on him like a wildcat. Blood ran from the prostrate bully's cheeks where Charley's caulks had bitten deep.

But the challenger got up. How, I don't know. Charley went down in the mad, bull-like rush that followed, and rolled clear. The big fellow then aimed a kick at Charley's groin, but Charley's foot hit the knee-cap of his antagonist, who dropped to the ground, writhing. Charley picked himself up slowly, shook the sweat from his eyes and gasped, "Take him to the hospital. Tell 'em he got hurt in line of duty." Then he turned to my father, who was standing beside me on the hotel porch, and said, "It was duty, too, John. I had to lick him."

Charley was later state forester of Montana. Now—well, I guess he's breaking jams on the river Styx, and arguing with Hiram, that other great logger, about how best to skid the cedar trees from Mount Olympus across the asphodel meadows.

The scene changes. I'm fourteen years old, and carrying two big panniers of lunch down the river trail to the men on the jam. Scents of green spruce timber and the odor of wild roses in my nostrils. A partridge cock drums once from a log. I approach a clearing, and hear the thud of body blows, the quick-drawn breaths of fighting men. In the clearing, bloody and battered, Ed Kocin and Red Jack Stringer, my idols on the drive crew, are battering each other unmercifully. I pause, fascinated, and set my panniers down. Around they mill—first a flurry of fists, then the sight of big bodies rolling on the ground. Red Jack on top, punching at Eddie's face. I let out a groan—unconsciously, I think. The two men stop fighting and come over to me.

"What's the matter?" I ask.

"Just a friendly argument, that's all, flunkey."

Something seems to stir the lumbermen's fighting blood when the rivers are muddy and the logs are floating free. If they can't find trouble among their own crew, they'll find it elsewhere. I was with a crew sacking the Flathead, on the right bank of the river, and one Saturday night was among a crowd of youngsters who went to a dance in a country school house situated on our side of the stream. The sound of the dance music carried far through the warm June night, and the strains must have reached the ears of the left-bank crew, for they were at the dance in full force.

Everything was fairly peaceable until our boys began to twit the left-bankers about the loss of certain cases of powder. Then Ken Masters, one of our gang, cut in on Lard Huggins, a left banker, and started to dance with Lard's girl. Lard swung on Ken, who was pretty handy with his fists, and the brawl was on. Two left-bankers started to help Lard put Ken away. Then all of our bunch joined in. Ken went into a corner, upsetting a lunch table as he did so, and held off two of the enemy for a minute or two. Three of the lads went to the floor and wallowed around in the lunch. (It's the only time I've ever seen a lovely white-frosted cake smeared into a logger's hair, but it was well-smeared that night!)

The fight waxed hotter, and the left-bankers were putting the run on us when the plow-trailers who were there decided to take a hand in the affair. They were going to help put us out on our respective ears, but all loggers know that when a clodhopper attempts to mix in a brawl between loggers, the interferer usually gets the worst of it. The loggers forgot their own dispute and began to wreak havoc on the country lads. One by one the farmers were tossed through a convenient window, while the girls stood on the sidelines, screaming like wild. All but one husky hay hand was thus dismissed; he put up a noble battle until four good lumberjacks pounced on him. One of our bright young men conceived the idea that the farmer had too many clothes on, so they stripped him down to his gorgeous red flannel underwear. The farmer's pants came off and were flung through a window. The girls screeched.

After the last plowboy had been ejected, we tossed the benches through the windows, and took the girls home. "To the victor belongs the spoils," and some of those girls were certainly spoiled beauties, all right!

Once, in a Coeur d'Alene saloon, I heard a cocky young river pig announce, "One man from Marble Creek can lick a dozen Coeur d'Alene high-bankers."

Coeur d'Alene loggers have ever been noted for their ability to handle any situation, and they took that cocky young fellow at his word. Close quarters were the order of the day. Three local men tied into the young cockrel and trimmed his spurs. He was hit a dozen times in the midriff and finally a bottle,, swung by a hand practiced in swinging a broad axe, crashed into his head. He slumped. When he came to, the brass foot rail of the bar was protecting his face from the caulks of a little red-headed wildcat. The caulks struck fire from the rail every time they hit. The lad who instigated the riot was taken to the hospital finally, and from what I can gather, has adopted less strenuous forms of amusement.

—From the 4L Lumber News

BIG WHEELS In The PINE

SLIP-TONGUE WHEELS used by Clover Valley Lumber Company, Loyalton, California. (Photo F. Hal Higgins Collection)

SLIP TONGUE WHEELS AT WEED. "Toggle knocker" has just released log and team is moving with slip tongue at longest extension. Photo taken by Charles R. Miller for McCloud River Lumber Co. about 1910. (Photo Miller Collection, Collier State Park Logging Museum)

ALL RIGHT ON LEVEL GROUND but no braking action on grades. Stinger tongue wheels brought from Michigan and Wisconsin and used by McCloud River Lumber Company. Note arched axle for big log. Tongue was hoisted with "johnson board", chain attached and team pulled tongue down, raising log. If log was not loaded heavy on front end, team had to outrun wheels. (Photo Miller Collection, Collier State Park Logging Museum)

STINGER TONGUE IN SUGAR PINE. Note dirt at front of wheel where hole was dug under log and chain dragged through with "canary". Wheels were usually chocked until log loaded. On these wheels horses weighed 1800 to 2000 pounds and were highly valuable. Note heavy britching. (Photo Parsons Collection, Collier State Park Logging Museum)

12 FOOT WHEELS AT BEND. Logs have been chained to hoisting rigging and lead team is ready to pull. These wheels were the slip-tongue type developed by Redding Iron Works. (Photo Paul Hosmer)

In the comparatively flat forest land of Oregon and California before the days of the trailer, high wheels served to get the short logs off the ground and horses skidded them to landings. They were a phenomena peculiar to the pine of Central and Southern Oregon and Northern California.

In the sugar and yellow pine, the McCloud River Lumber Co., successor to Van Arsdale and Harris, first used the "stinger tongue" wheels which had been brought out from Michigan and Wisconsin where the ground was comparatively flat. Some were built with arched axle for big logs. They were dangerous rigs on steep shows as they had no braking facility. Men who logged with them say "it was worth your life to drive a team in front of a stinger tongue load thundering down a grade."

This handicap gave rise to the "slip tongue" wheels or carts, developed by the Redding Iron Works at Redding, California. This type was used at McCloud River, Shevlin-Hixon at Bend and other operations. They were 12 foot wheels, with iron rims, pulled by 4-horse teams. The wheels had a 6-foot gauge, were connected by a heavy axle with sliding tongue which operated a hoisting device to raise logs off the ground. Of the four horses, two were in the lead, two on the tongue—drivers riding on one of the wheelers.

A crew of men bunched the logs, and made them up into loads, a gopher man digging a shallow trench under the load just forward of the weight center. The teamster backed the wheels astride the load and the logs were chained to the hoisting rigging. As the team started ahead, the tongue slid forward, pulled on a chain and the forward ends of logs were lifted. The harder the team pulled the higher the logs went. Down heavy grades, the tongue was pushed back to let logs down and act as brakes.

At landings, the logs were unchained and loaded on railroad cars by a jammer, the wheels being returned for another load. Shevlin-Hixon used this method for about seven years, during which time it took out 500 million feet of pine—91 million in 1922 alone. This era ended when arches came in. The big wheels were scrapped, their last utility serving fallers who cut up the oak spokes for falling wedges.

CAT 30 IN OREGON PINE just before arches made high wheels obsolete. (Photo
F. Hal Higgins Collection)

CAT 60 AND IRON WHEELS at Big River, California redwoods in '20s. (Photo
Union Lumber Company Collection)

HIGH WHEEL GRAVEYARD at Bend. Wheels went out when arches and bummers came in. Fallers cut out hardwood spokes and made them into wedges. (Photo Paul Hosmer)

TWO H.P. WOOD BUCK TREADLE, powered by a pair of live horses, cuts fuel for donkey on Fisk logging operation near Acme (now Cushman), Oregon, on Suislaw River. Photographed in 1900 by Phil Nicolle. (Photo courtesy of Mrs. Carrol T. Hays and A. A. Lausmann)

HUMP - -YOU BUCK

Alfred D. Collier

Getting around to the subject of bull teams, they tell a story about Bill Tingley, when he was top bull driver for Abner Weed. He got a big load of logs stuck in a mud hole one time, right in front of the saloon. While the oxen rested he went in and got a couple of snorts. Outside again, he started with the wheel yoke, went up the line touching the bulls in back of the front shoulders so that the whole team hung in the collar, meanwhile exhorting them with choice language culled from the book of life. When the goad and the words failed to move the load, Bill jumped up on the back of the lead bull with his caulked shoes and started treading out the forerunner of the square dance in plain sight of the rest of the team. The agonized bellows resulting supplied the necessary incentive to the rest of the bulls and out came the load.

A bull was named by the driver either for coloring or peculiarities or maybe for the owner from whom he had been bought. At the discretion of the driver he might be named for the person whom he particularly wished to cuss out for his shortcomings. This was a real economy as it permitted the driver to tell everyone within the sound of his voice exactly what he thought of the person involved without having to indulge in bodily combat and resulted in a lot of work being done by the bulls because of personal reasons. When they worked on the skid road the bulls learned pronto to run for dear life whenever the logs quit dragging. The wheel bulls pulled out as far as they could for if the hoof got between the skid and the log, there'd be ox meat on the table.

The Oregon law limited the depth to which you could stick the ox goad in, but the drivers would take the washers off when they got into a hard pull. When the oxen couldn't pull the load and the driver wouldn't give up, the bellowings of the bulls would make even a strong man cringe.

A Paul Bunyan yarn comes to mind—about the time Old Babe got the warble worms in his back and just couldn't keep his mind on his work through the itching. After he had rubbed out about an acre of nice sugar pine, Paul decided that something simply had to be done about the situation. He figured he would first have to pare the hair off old Babe's back. He sent east for a Bessemer steel razor blade, oil hardened and tempered, and lathered up old Babe's back. But alas and alack! Babe had been around those lumberjacks so long his hair was really little horns and the steel blade slid right over the top of them. Undaunted, Old Paul hunted up a thrifty bull pine and peeled her down till he came to the place where they had had the two summer's wood and laid it out in the sun to dry. When it got good and stiff, he just took the dishwater and sloshed it on and with that stiff timber just peeled those hairs off like they were sprouts on a redwood burl. When he got through he had such a nice slick place on old Babe's back that the ox lost half a day just doing nothing but licking the shaved spot. Then Paul tried to get those warble worms out every way that he knew. It was nice and warm under Babe's hide and those worms didn't want to come out. But a warble worm is a hot blooded animal and so he finally got Shanty Boy up there to play his squeeze box so that

ON STILLAGUAMISH IN 1893. Skid road team at Pete McMartin's logging camp.
(Darius Kinsey photo from Jesse E. Ebert)

everybody got to jumping and some of them landed on this smooth piece of hide. Well all this started to tickle Babe and he thought it was flies and jerked that skin this way and that. Now these warble worms got far out to the edge where it was real slick and they just couldn't get back so they fell off and hit the ground so hard they went right down into it. And the ground kind of closed up so they couldn't get back. Believe me, those warble worms were so darn mad they began to sputter and even today they're still sputtering. Some people say it's the hot springs coming up but loggers know better—it's those warble worms. Needless to say they didn't bother old Babe any more.

Here's something interesting about ox yokes. Most of the old yokes were rather hurriedly made when one was broken or worn out and very little time was wasted in the shaping of the wood for it. Generally they chose a piece of wood near the butt of the tree because of its greater strength. Normally they chopped as much as they could and if they had a spoke shave, smoothed it off so that it would not rub the ox raw. They might even put a little axle grease on it so that it would slip a little easier on the ox as he put one foot ahead of the other.

BULL TEAMS IN THE SEQUOIAS

A young girl's experiences in the early logging days are recounted in the Tulare County Historical Society Bulletin of October, 1950. At a picnic at Big Stumps, Mrs. Lizzie McGee read a paper about her pioneering girlhood. Her skidroad description:

"Bull teams, hauling logs from the woods to the mill over skidroads, picturesquely dot the hill side. A bull team needs no harness. A yoke fits across the necks of each pair of bulls. This is hooked to a heavy chain that in turn is hooked to a big log, or sometimes several of them stretched out. The bulls are pretty securely tied together and lift the load with their strong necks and shoulders. A span of six, eight or ten bulls represent a powerful lift. Bill McGee drove one team. He had Bright, Brigham, Buck, Brin, Star and Hank and others. With a sharp goad stick he commanded obedience. He got them in motion with a light jab on the rump of each one. They began to lean forward, backs humped, the yokes began to creak, the chains clinked and the logging chain straightened out. The bull whacker kept alert. If an animal didn't take a step when the rest did he got a good punch with the goad. It reached up, over and down on the rear of Mr. Bull. If the team was too reluctant to get in motion McGee managed to get in some quick action. In rapid succession he jabbed heavily each bull's rump and emphasized the jab with a stout swear word. They moved evenly into pulling strength. In the morning Bill's voice rang out loud and clear, "Gee! Henry, Bright, Brigham, Buck, Brin and Star!" with his pet 'tuning up' booster oaths that fascinated us children and horrified some of the more serious elders. By noon the volume of his vocal output was considerabily quieted. By quitting time there was only a croaking gutteral tone. The logs were hauled to the mill over skid roads. These were constructed of eighteen or twenty-four inch thick logs half buried in the ground. Coming down inclines they were laid lengthwise. A skid greaser and swamper accompanied each team. They swabbed on skid grease where going was tough. Bill McGee was killed when a log jumped sidewise and pinned him to a stump."

LOGS AND SILK DRESSES—Bull team hauls big load on solid wheel wagon in Eastern Oregon, about 1885. (Photo Caterpillar Tractor Co. from F. Hal Higgins Collection)

VOICES FROM THE DIM AND DISTANT

Alfred D. Collier, curator of Oregon's Collier State Park Logging Museum, and a former logger himself, recounts conversations with two colorful old timers—James R. McCrank and George A. Campo.

"They were old-school loggers," Cap Collier states, "who insisted on doing everything for themselves—by hand, the hard way we think now. They worked on the Klamath River drives and when they came to town the newspaper editor would say, 'Here comes those two amphibious swine.' Jim McCrank was still alive in '52, still using kerosene lamps and as independent as a hog on ice.

"Jim once told me a chap by the name of Truitt had a saloon at the foot of the old Pokegama chute opposite Shovel Creek, a spot called Old Snow. Old McDonald, a saloon keeper at Klamathon, came up to collect a bill with two girls in an old buckboard. Henry Hoover, McSherry and another fellow took McDonald's buckboard and started to drive off. McDonald told them to stop. They kept on and McDonald shot and killed Hoover. Sheriff Ed Laux tried to arrest McDonald and he too was shot. McDonald was sentenced to prison for this and Jim found him years later quite prosperous after his stay in prison. He warned Jim to leave him alone or he would kill him.

LOGGERS LOAD OUT. Two ox teams ready to be chained to logs in deep woods. (Photo Oregon Collection, University of Oregon)

"This story is corroborated by Jim White, 82 years old and still riding for Hessigs near Shovel Creek. He says that McDonald had two girls in the buckboard and that Hoover got in between them, told Hoover to get out and he didn't, so he shot him. Hoover was a brother of Bill Hoover who is still living above Shovel Creek on the other side of the river. Jim said later that H. Harvey Edmonds was head man for the Algoma Mill at Spring or Eagle Creek and that Ed Way got too much interested in his wife; that Edmonds invited him over to look at a new shotgun and when he came to the door, kicked him in the groin, and jabbing the muzzle of the shotgun in his mouth, shot him. Then Edmonds and his wife took Ed Way's Ford car and started for town to give himself up. The car stalled on a crooked hill and he walked on in to town. He pleaded the unwritten law, his wife supported him in it and he went scot free.

"Jim McCrank said Wilbur Hale was the head push and Jack Villair drove wheels for them. Jack later married Mable Geisberger who was a clerk for Goldwaits and had brains. Carl Lang writes of a similar situation in his 'Women and Yamsay Loggers.' He recalled Ferguson Brothers, in Michigan, had an upright boiler on a sleigh with pipes for runners and holes in the nose. The first team out in the morning when the frost was hard, would blow steam in the ruts and the steam would shine and they could haul a load up to about 20,000 pounds. The runners would freeze down to the ice so they had to use a rock and pinch bar under each runner, one man under each runner to start the lead. He said they used housings on the horses' col-

MEN OF YEON. Mustaches everywhere as John Yeon's loggers are photographed in front of bunk house. (Photo Oregon Collection, University of Oregon)

lars to keep the snow and rain off their necks where the collars worked on them, also dust. When they shod the oxen they pulled the leg out backwards and made the ox fast.

"Pope and Talbot at Port Gamble on Puget Sound used oxen for two years. For a buffing bar in loading they used a single tree with a spud in each end, the spuds turned the opposite way. When it rained heavy they would say 'it was raining pitchforks and saw logs.' When they came to a steep place on their skid roads they would put a yoke or two of oxen behind to hold back. They also put a pole into the yoke of the first team to crowd the oxen ahead of the log when it picked up momentum so that they would not cut off the heels of the oxen. At the head of the chute they used oxen to start the logs, knocking the hooks out with a post.

"The old loggers up in Washington tell of the time they were trying to move a donkey across a creek. The line parted and the donkey fell backward into the water. They had no steam but they had to get the donkey out, so they took another big line out to the donkey, dropped timber on it to shorten it and managed to get the donkey out. To increase the purchase on the loading line at cross haul, they would put a 'sampson' or 'bitch block' under it. Sometimes they put a stick out in front to increase the lift so that the log would go up and over a windfall.

"SMILE—YOU BRUTES!" Bensen-Yeon driver stopped this team for picture. Note yoke details. (Prentiss photo Oregon Collection, University of Oregon)

"Jim McCrank said that in Washington, steam for logging came off the ships as the Finlanders and the sailors were used to this idea and brought the steam in. He tended spool on the Sound for an old donkey in the 1890's. They used a line of two inches in diameter or over and a horse for the haul-back. It took three or four turns on the spool to start. The old rope would stretch and zing and the log would jump when it started. They never went over half a mile. To pile up the slack they used a figure 8. When the log hung up on a stump they would let the rope slip and then work it around the stump. Most of the timber was barked so it would slip easier and most of the logging was done on slopes adjacent to salt water.

"Then there was 'Walkaway Burns', who got his name by walking away from the jobs that didn't suit him. Once at the huckleberry patch, he asked Holly Miller, an Indian, 'Have you got anything yet?' She answered, 'Yes, and huckleberries too'. About the first mention Jim McCrank makes of Paul Bunyan on the West Coast was the time that Seaton, the cross haul man, went to town and didn't get back. The push was angry and said, 'What in hell kept you there?' Seaton replied, 'Oh, I had to keep old Paul company,' whereupon the push fired him. Jim told about a gyppo logger who was behind in his bills and came running back into camp with the story—'The sheriff is coming. I don't have money to pay you off, so everybody grab something and the big man take the grindstone.'"

"Jim and George Campo talked about the traps they used to make in the old time logging camps. They would cut a V in the top of a high stump and bait it with a grouse. When the wolf or coyote tried to get the grouse and started to pull down his paws, they would catch in the V and keep on tightening the harder he pulled. They also talked about a barrel trap for bears. They would take a heavy oak beer keg and drive spikes inward near the bottom, putting honey in the bottom of the keg. The bear would poke his head into the honey and when he tried to pull it out, the nails stuck in his neck and held him.

"Ray Chase tells us of a sport they had at Shovel Creek on the Klamath River at the bottom of the Pokegama Chute. They would knock the end out of a beer barrel and put a burlap sack over it, running a 3/4-inch rope through the sack. A new man would be told that there was a badger in there and they would bet $10 he couldn't pull him out. The newcomer generally bit. He would take hold of the rope and brace himself for a tremendous pull and when the rope came out freely the fellow would go flying on his heels. The drinks were then on him. The reason this rope trick was a sucker game was that occasionally they did put a badger in the barrel and would bet that no dog could get him out for $10. This was true because the badger, being protected, could drive the dogs out."

TUNNEL ON WESTPORT SKIDROAD. Possibly unique in West Coast logging was this tunnel of the Westport Lumber Co., near Westport, Oregon, in the early 1890s, where enterprising loggers bored through hill rather than lay an extra half-mile of skidroad. Bullteam is returning to woods after hauling turn to landing. Tunnel is still to be seen near highway between Portland and Astoria. (Photo Stewart Holbrook Collection)

BULLWHACKER GETS POWER in team of ten oxen on Bensen-Yeon skidroad in Oregon. Skid greaser at left changes buckets. (Photo Oregon Collection, University of Oregon)

LANDING IN GRAYS HARBOR. Note skidroad on slope at left. (Photo University of Washington)

OREGON

High production had not yet hit the woods. It was a day of catch-as-catch-can and get the logs out the handiest way. Bulls and horses were good enough to drag them a mile or two and there were the rivers to get them the rest of the way to the sawmills.

That was the way it looked in the pine country of Oregon around the turn of the century. And all the colorful details of the actual river drives are recounted faithfully and "fun-fully" by H. J. Cox in his book, "Random Lengths". Herb Cox was a logger on the business side and passed through many phases of pine logging with an observant and objective eye. It is possible to imagine a trace of mistiness in that eye as he wrote about Oregon river logging.

"River loggers," he says, "considered themselves the peers of the woods loggers and referred to them as 'timber beasts', whose job of cutting down trees and dragging the logs out of the woods was inconsequential as compared to the ingenuity, perseverence and endurance in overcoming the moods and vicissitudes of Ol' Man River.

RIVER DRIVE ON WILLAMETTE near Jasper, Lane County, Oregon, by Booth-Kelly Lumber Company in 1905. (Photo courtesy H. J. Cox)

RIVER DRIVE COOK TENT on Fall Creek—Booth-Kelly Lumber Company, 1905. (Photo courtesy H. J. Cox)

BOOTH-KELLY CREW, mouth of Big Fall Creek at confluence with middle Willamette River, 1905. (Photo courtesy H. J. Cox)

FIRST STEAM ON MILL CREEK. Booth-Kelly Lumber Co. donkey at Wendling, Oregon, 1901. (Photo Oscar R. Page Collection from Collier State Park Logging Museum)

BOOTH-KELLY MEN—1901. Left to right—Art Thompson, H. R. Ross, C. Taylor, Hart, Oscar Page, Jim Doley and C. Parks. (Photo Oscar R. Page Collection from Collier State Park Logging Museum)

"The boss of the woods, unlike modern scientific logging bosses, many of whom play golf and hang around country club bars, did not depend on, or need, or burden himself with any helper army of foresters, logging engineers, cruisers, tree markers, and other fern hoppers, all rushing hither and yon, up-hill and down, carrying barometers, binoculars, microscopes, surveying instruments, thermometers, relief maps, bug killers and test tubes to get the logs out of the woods and on the way to the saw mill.

"All river crews, especially where the company owned and operated logging camps and drove several streams, had a river boss for each driving crew, and each boss was discussed and cussed by a gentleman who carried the title of general logging superintendent. This individual rarely showed up on the job, and on such few occasions, he did not get in or go near the water, being clad in store clothes, white dickie and celluloid collar, and a permanently tied bow tie fastened around his neck with an elastic hooked band. On his head he wore a 'hard hat' (derby), and on his feet a pair of 'hen-skin' buttoned shoes. He was fated to appear on the job at the time the river boss was having a bad time over breaking up a log jam or working stranded logs off a tough gravel bar. Running back and forth along the bank, he would wave his arms and unintelligibly shout and rant at the river boss; the latter paying no heed, knowing the 'super' was not garbed or shod to enter water or walk on logs.

"The river boss, as well as the woods boss, did not have to consult some labor union business agent in order to hire or fire a man, and the man who held up his end of the job was never fired. Nor did the boss reach his position through influence or inheritance. Often, to prove his fitness and command respect of his men, it was necessary to whip the camp bully, as well as any other member of the crew who undertook to question his authority. On the other hand, the boss would not allow a crew member to expose himself to any danger in which he himself would not take the lead. He was the first to come to the financial aid of a worthy workman, the last to start a fight. On the other hand he would rush to the defense of his men, involved singly or collectively, unintentially or otherwise, in a brawl with city or bar-room rowdies.

"PASTOR OF THE PINES" and biggest Ponderosa pine on earth. With trailer, family and bible, G. O. Redden traveled over the pine country taking messages of inspiration to loggers. (Photo Jim Hosmer)

"River loggers were of two species, the local home-guard boys, and the 'river rats' who migrated from state to state and from one log drive to another—a type comparable to the present-day itinerant fruit picker, except that a 'river rat' either had no marital status, or his wife, and children, if any, were not dragged around the country with him.

"When the boss of the woods had sufficient logs in water, the time had arrived for the river logger to step into his role of professional extrovert. When logging was done in smaller streams tributary to the main river, it was necessary to construct a series of 'splash' dams, impounding the water in the upper dam, then releasing and flooding the logs through one dam to the next until the logs reached the main river. Floating boom timbers tied together with wire ropes and fastened to a steel cable attached to each bank of the main river temporarily held the logs as they arrived from upstream. Boom men, using long pike poles, diverted the logs to the millrace where they floated with the current of the stream into the mill pond. River log driving started after the danger of spring floods was over, and if a fall drive, before the danger of winter high water.

"As the thousands of logs came floating down the stream, many of them would lodge on river gravel bars, in shallow water, and in many instances, a moving log would hang up against a protruding rock or other river obstruction, causing the fast moving rear logs to pile up and cause a 'log jam'. This was where the river logger entered the picture in all his power and glory.

FLUME FROM MILL TO RAILROAD carrying sawed fir lumber from Booth-Kelly mill in hills to Saginaw, Oregon, around 1900. Flume 110 feet at highest point. Men sometimes rode big timbers or bunched pieces. Loads occasionally jammed and some lumber had to be thrown over side. See sticks at foot of trestle. (Photo Yancey Collection, Collier State Park Logging Museum)

"Preparations for a log drive were carried out with pomp and ceremony comparable to an Elks Lodge Fourth of July picnic. Horses were reshod with river-caulk shoes, iron horse shoes with sharp pointed pieces of metal projecting downward on the front and rear ends of the shoe to prevent slipping on bed-rock or gravel bottom. Harness was checked, repaired and oiled. Peavies, stout near the spike end, were assembled. Chain dogs and chain, dogger mauls, and tripper bars were made ready. Blanket and cook boats were put in usable shape.

"While their preparations were under way, the bull cook, affectionately known as 'camp louse', was busily engaged in assembling hay and oats for the horses, cook stove, camp dishes, pots, pans and kettles, together with a supply of foodstuffs from a list submitted by the camp cook, and including beans, bacon, spuds, hams, flour, canned fruit of all kinds of the best brands, coffee, spice, vanilla extract (the brand containing 50 per cent alcohol, and for emergency use), and all other condiments and trimmings necessary to make the mostest and bestest fed branch of the forest products industry.

"When ready, the river boss, having relayed word to all crew members with the assistance of the camp louse, everything except the river horses was loaded on freighting wagons and conveyed to the starting point of the log drive, usually 30 to 40 miles or more upriver from the sawmill. The river horses were taken under their own power to the place of beginning. The crew reached the starting camp by various means of transportation, some riding on the freighting wagons with their individual bed roll of blankets, underwear at least one-eighth inch thick, an ample supply of long woolen socks, the supply being as large but not of the same texture as those given away on the radio 'queen for a day' programs of today. Also, his complement included a pair of logger boots, fully caulked—around 12 inches tall with thick soles to hold steel caulks, and heavily reinforced to protect the ankles and achilles tendons. The primary purpose of steel caulks was to prevent slipping on logs or river bed, likewise, an effective weapon for offense and defense.

"While loggers required no artificial appetizer, the large majority arrived in camp with as much liquor as they could carry, both inside and out, or as much as the freighting wagon space could accommodate. Those tarrying too late in town arrived by livery rig, well loaded with Shaw's Malt, a whiskey then as popular among loggers as Borden's Milk was among infants.

"The river crew and equipment consisted of one feed and blanket boat, one cook boat, one cook, one cookee (who also served as boatman), one bull cook, twelve rollers (men who rolled the logs with peavies), six to eight teams of horses, with one driver for each team, who rode astride one of the horses, the horses furnishing the motive power to roll the logs and drag them to deep water where the river current would convey them downstream. Three doggers, each with an iron maul, drove the iron dogs into each log — the dogs being attached to one end of a logging chain, a horse team hitched to the other end. The chain was adjusted on the log as to cause it to roll toward deeper water with the assistance of as many 'rollers' as necessary to complete the maneuver. Three 'trippers', each of them riding an individual log to deep water, released the dogs with an iron bar as the horse team swung the log into the deeper current, then jumping from the log into sometimes armpit-deep water, wading out and mounting the next approaching log.

"Through all this, it was up to the river boss to do the heavy thinking, especially when it came to breaking a log jam. He was the one who figured where and how the key log would release the jam, or he jumped from log to log, carrying dynamite in an old gunny sack, which he planted at the vital spot, then rapidly retracing his way over a jumble of logs to reach the river bank before the explosion released a roaring sea of logs which tumbled end over end on their mad rush down river.

"As each bar was cleared of stranded logs, or the log jam broken, the entire operating crew, with horse teams and tools, would move downstream to the next log jam or other river obstruction. Bringing up the rear from day to day were the feed, blanket and cook boats; the feed and cook boats catching up with the crew for a gravel bar or river bank noonday meal, then going on downstream preceded by the blanket boat, and making camp where the crew would quit at the end of the day. This procedure would be repeated daily until the river crew followed the last log to the river log boom at the inlet to the mill race, at which place the boom crew took over, and where for several

HALF MILE OF JUNK AND ASHES. Smouldering remains of Hammond Lumber Company's Astoria plant, completely destroyed by fire September 11, 1922. (Photo Oregon Collection, University of Oregon)

weeks they had headed the 6 to 12 million feet (log scale), on its final lap to the sawmill log pond.

"Unless the river crew had been running a two-company drive, such as Booth-Kelly and Spaulding on the McKenzie River, their 30 to 40 mile, 30 to 50 day job was done; but in case of the latter, they would continue on, driving the Spaulding logs which had been passed through the Coburg log boom, on down the McKenzie River to its confluence with the Willamette, and from there, an additional 75 to 80 miles, 75 to 100 day trip to sawmills at Salem and Newberg.

"A few miles below the mouth of the McKenzie River (Leach's Bar), and the Willamette, a river of deeper depth, the crew, horses, food stores and all equipment would be loaded on wangans, large log rafts, floored with plank lumber and containing a limited amount of superstructure. This flotilla was made of two horse wangans, each holding twelve animals; one cook wangan with dining quarters, and one wangan for crew sleeping accommodations. The flotilla would follow the river crew downstream, tying up for the night along the river bank, preferably, and if they could possibly make it, near one of the small river towns, thus permitting crew members, so inclined, to go ashore in search of liquid refreshments and such.

"To augment the original food and horse feed stores, it was one of the duties of the camp louse to drive to the nearest town and replenish food stores according to the shortage list submitted by the cook. Spuds, fresh meat, poultry, eggs, vegetables, and horse feed were purchased from farmers living along the river, who were given written orders on the company which they cashed at the company's general office. Also, the camp louse was required, upon individual request, to replenish the liquor supply of the applicant, which was usually obtained from Shumate's General Merchandise Store at Walterville, as Walter carried the largest stock of Shaw's Malt, as well as being able to read and write, thus assisting the camp louse in getting the accounts turned in to the company, where they were deducted from the loggers' pay at the end of the log drive.

"The extent of a camp cook's prestige, commendation and length of service was governed by his culinary ability and his personal hygiene. There was an instance of one cook who did not get his name on the payroll or in the tablet of industry fame. This cook was the possessor of a glass eye, which at night he placed in an empty coffee cup on the dish cabinet shelf so as not to forget replacing said eye in its socket before starting breakfast. The river boss, arising during the night for purposes, one of which was to settle a stomach revolting from early evening doses of Shaw's Malt, reached for a cup with one trembling hand, and with the other grabbed the coffee pot still simmering on the stove. Gulping the contents, the bottom of the cup quickly appeared —with an eye staring him squarely in the face. Breakfast was prepared and served by the cookee and camp louse, assisted by the river boss, who before daybreak had forcibly headed the cook on his way toward new pastures.

"Two outstanding cooks were Pug Huntley on the McKenzie and Jim Awbrey on the Williamette. Pug was one of the best known cooks in the history of river cook house mechanics. His specialty was pastry, as well as other

WATER POWER AT THE DALLES. Manchester & Lester water power sawmill on Five Mile Creek about 1908. (Photo G. E. Manchester)

FALLING WITH RUBBER MAN in Brooks-Scanlon camp at Bend. Iron stake was driven in ground, strip of inner tube fastened to it and saw. Faller pulled saw, Rubber Man pulled it back. Ideal for fallers who preferred extra wages to company and back talk. (Photo Paul Hosmer)

tasty and nutritious dishes to which a river logger had been unaccustomed, either at home or elsewhere. Slightly on the debit side was Pug's tendency to overlook his professional personal appearance. He would start on the river drive with a cook hat and apron as clean and white as a virgin's heart, but through oversight the raiment would soon resemble Jacob's coat. When this eventuated, a delegation from the crew would wait upon Pug, pressuring him to dress up. During the balance of the drive, Pug would keep within the law."

Herb Cox describes the river operations of the Booth-Kelly Lumber Company in rich detail. The firm started logging in the summer of 1902 and in March, the following year inaugurated its first Fall Creek river drive.

"The camp was on the Matteson place between Big and Little Fall Creek. The ranch house was converted into a cook shack. A tent was used for bunk quarters and accommodated about 20 men. A chicken house was also used for sleeping purposes. The tent was heated by a barrel stove. Occupants furnished their own fuel which consisted of bark chunks, limbs and other debris picked up on their way in from work at night. Such a thing as a 'bull cook' was unheard of in those days. Bunks were of the double deck variety and constructed of inch lumber. The mattresses were filled with hay which was always plentiful since logging operations were carried on by horses. Light was furnished by means of a small kerosene lamp which was about equal in candle power to an average parlor match. The logging superintendent for the company in those days was a man named L. S. 'Bunker' Hill, assisted by a foreman named Wallace Warner."

FAMOUS KNOT TREE. Over-production of burls on spruce tree, called Neh-Koh-Nie by Indians near Nehalem. (Photo Oregon Collection, University of Oregon)

Oh Stranger, ponder well, what breed of men were these
Cruisers, Fallers, Skinners, Ox, Horse, and "Cat"
Chokersetters and the rest who used these tools.
No summer's searing dust could parch their souls,
Nor bitter breath of winter chill their hearts.
'Twas never said "They worked for pay alone"
Tho it was good and always freely spent.
Tough jobs to lick they welcomed with each day,
"We'll bury that old mill in logs," their boast,
Such men as these have made this country great.
Beyond the grasp of smaller, meaner men.
Pray God, Oh Stranger, others yet be born
Worthy as they to wear a Logger's Boots.

— "Ode To the Loggers" — By Nelson Reed

The 146 acres of Collier State Park proper represent a gift to the state of Oregon from Andrew and Alfred Collier in honor of their parents, Charles Morse Collier, pioneer U. S. engineer and surveyor, for Lane County, and Janet M. Collier, Daughter of Pioneers of 1853, a school teacher active in promoting the University of Oregon and the Eugene city schools.

The setting on U.S. Highway 97 at the junction of Williamson River and Spring Creek contains some of the original virgin yellow pine timber and a good reproductive stand of pine. The State of Oregon built and maintains the roads, supplied landscaping and caretaker. The museum buildings and the equipment have been donated by individuals. Alfred D. Collier has been in charge of planning and solicitations with assistance from Sam Boardman, former Park Superintendent for the Highway Commission, now deceased, and his successor, A. H. Armstrong.

The log buildings of the museum are covered with extra thick sugar pine shakes. Inside are ox yokes, several shoes, the stocks in which bulls were trussed and strung up for shoeing; solid ox-drawn wooden wheeled wagon and a collection of pictures and stories of bull teams and bulls in combination with horses; four horse, spoke-wheeled wagon; high wheels, both stinger tongue and slip tongue types; sleds used on snow, varying from the crude flat bottomed home made type to the oval shoe McClaren Logging Sleigh, two of the latter loaded with large fir and sugar pine logs. There is a representative showing of equipment used with horses—collars, housings, stretchers, neckyokes, equalizers, shoes. The blacksmith shop exhibits old leather bellows, hand blowers, forge, tire shrinkers, anvils, hand tools, funnel or cone for making perfect circles.

In the steam power section there is Case Engine No. 319 made in 1878, used in a sawmill and as a road donkey; Tacoma yarder; wagons used behind steam engines from solid wooden wheels to 30 ton Reliance steel wheels with Timken bearings; Baldwin locomotive. Gas engines are represented with the first Holt 22 caterpillar and set of high wheels; Best 30 and 60 Cat, all in working shape and ready to run; the first Best arch, Glasscock slip tongue arch and Athey arch. Displayed also is a Mack chain drive truck.

POKEGAMA CHUTE INTO KLAMATH RIVER. Logs smoked and burned as they shot down 2650 foot slide at 100 miles per hour. (Photo Ogle-Baldwin Collection, Collier State Park Logging Museum)

THE POKEGAMA CHUTE

It must have been a sight for even those sore-muscled men—big peeled logs streaking and smoking down the greased track like bolts flung at the hills by the gods. This was the Pokegama Chute operated in the yellow pine of southwest Klamath County, Oregon, in the 1890's, the phenomenal 2,650 foot slide into the Klamath River.

The problem the loggers had faced was how to get the timber out. It stood in mountainous country, and a logging railroad from timber to mainline was unheard of in those early days. Yet this certain tract was only 30 miles from the Oregon and California railroad river crossing at Klamathon.

So John Cook built a mill at Klamathon in 1892. At the timber end logging headquarters were set up at Snow and a narrow gauge railroad built down grade from camp to a high spot on the Klamath River canyon. At that point the big chute was constructed.

The Pokegama Chute used small logs for the bottom, large ones to form the sides and the bearing surfaces were hewn smooth. Sections of the chute lay flat on the ground while others were carried over gullies with low trestles. At one point the chute passed through a deep cut.

To bring the log cars back from the head of the chute, a narrow-gauge locomotive was brought in, the first in Klamath County. John Cook engineered the transporting of it with a team of ten bulls. The engine was a Baldwin and nicknamed "Old Blue." It had a run of about seven miles and the fuel slabs had to be less than five inches in diameter to get into the fire box.

The log cars were hand braked down to the chute head where the logs were decked and cant hooked into the chute. Logs were peeled to slide easier, so they would be less liable to catch fire or jump out of the chute. They were started down by a "Walking Dudley"—the vertical spool donkey. The men called the cant hooks "log wrenches" and as they bent with use, the expression "throw a sag into her" grew common. No carpenter was more fond of a plane or chisel. Some of the Pokegama loggers

LOGS ON DECK AT HEAD OF POKEGAMA CHUTE. Note sag in cant hooks from which came expression, "throw a sag in her." (Photo Wren-Frain Collection, Collier State Park Logging Museum)

practiced with the log wrenches until they could catch a log in almost any position, stop and change its roll. Some men took their cant hooks with their packs to new jobs, rubbing the handles fondly with linseed oil and hanging them up at night.

The full downward sweep on the 2,650 foot chute put the smoking, sometimes burning logs into the river with a 90 foot plume of water, in 18-20 seconds–a drop of 834 feet. On the upper chute skid grease was daubed on the hewn timbers to make logs slide easier, on the flat lower section diamond shaped spikes were driven in to slow their progress. Some logs split when they hit the river and on frosty mornings the logs often shot full across.

At the river end a watchman used flags to signal logs down or hold them up–flag up, no logs; flag down, freeway. Occasionally the logs jammed and all hands assembled to clear the chute. In the river, logs were rolled into side channels until enough were accumulated for a drive down to the Klamathon mill.

Until 1902 the system worked well. John Cook who owned the sawmill had been financially ruined by the 1893 panic and in 1897 had leased the mill to Hervey Lindley. There was a disagreement, subsequent suit and Cook obtained a judgment of $40,000 in 1902. Later that year the mill and box factory burned with a half million dollar loss and there was further trouble with the miners who had wing dams on the gravel bars which ended in an injunction against the timber company's use of splash dams. This ended the river operation and brought on the building of the Klamath Lake Railroad up the river, later abandoned when the Southern Pacific built through the Cascades.

The Klamath Lake railroad served a mill built at "old" Pokegama by the Algoma Lumber Company who later moved up on upper Klamath Lake near Section 37. Mason, Lindley and Coffin also built a mill that shipped their lumber out over the Klamath Lake Railroad. When Harriman decided to build from Eugene through Klamath Falls to Weed, the Klamath Lake Railroad was no longer the answer to getting out the timber in the Klamath area and so was discontinued.

RIVER CREW AT FOOT OF CHUTE herding yellow pine logs after smoking ride down chute. (Photo Ogle-Baldwin Collection, Collier State Park Logging Museum)

COLOR AND CHOLER

No better champion of the working logger can be found than H. J. Cox, author of "Random Lengths", an informal presentation of his own forty years in the industry. He has the wholesome viewpoint that, good or bad, every logger had his usefulness and somewhere in the telling there is always a chuckle.

Racy Matteson was Booth-Kelly woods boss at Saginaw, says Herb, and if Hollywood had been in existence would surely have out-shone lesser stars. He leaned toward oratory and using a tree stump would render sermons to his men, complete with invocation and benediction. He read law in his spare time and kept many a logger out of trouble's clutches.

The only doctor in the mill town of Coburg was Milton Emerson Jarnigan, a native of Tennessee, under contract to the company. He never refused to make a call, never sent a bill. He loved the life and wrote glowing letters home inducing them to come to Coburg. One stranger who had trouble getting a physic of Doc's to work, finally told him he was from Tennessee. "My Gawd!" the good medic exclaimed. "Why didn't you-all say so? Here's four bits. Go get a good meal."

WALKING DUDLEY—DONKEY ON WHEELS employed by Henry Colvin, Marshfield in 1903. Same type used at Pokegama. (Photo Caterpillar Tractor Company from F. Hal Higgins Collection)

"OLD BLUE" BALDWIN—first locomotive in Klamath County, hauled in with ten bulls by John Cook in 1891.

Stalwart logger Cliff Abrams came from Crawfordsville, Oregon, earlier known as "Tail Holt". His father was a bull-team logger and in 1895 took a contract with J. H. Edwards Lumber Co. on Chandler Mountain which gave Cliff his first job. Then he drove team and scaled logs for Booth-Kelly at Matteson, in 1903 followed the first B-K log drive out of Fall Creek.

On these drives there was a continual turn-over of cooks since they had to do everything in most primitive conditions and a bullcook was unheard of. Tents were bunk quarters, heated by barrel stoves and the loggers scavenged their own fuel. The cook's hot water tank was a wooden barrel piped to the range. One cook was big Ed McCormack who arrived with practically nothing but old french-heeled logging boots. He had trouble keeping water from burning but worked twenty-four hours a day until at the end of three weeks he quit, utterly exhausted. "Them sons of bitches want cake and pie and a nigger to stand around and feed 'em!"

Jack Magladry was a human dynamo—all 120 pounds, 5 feet height of him. He came from California to Washington and after being "camp louse" at a small horse-logging outfit, was head sawyer at Yesler's Mill in Seattle. He was as scrappy as they came, could whip his weight in wildcats. He claimed to have been shot in the mouth in a barroom brawl, spitting out the bullet. At 60, when his teeth gave him trouble, part of the lead was found pressing against the nerve. Jack became shift sawyer at Inman-Poulsen in Portland, later invented "Magladry's Dream", a complicated road engine which went into flying parts on the first trial. With Con Keller he operated the Humes logging interests at Tongue Point on the Columbia and then with the Kellys, purchased the Donna plant on the Mohawk.

TIMBERS FOR TRADITION. One of five carloads of Douglas fir for masts and spars in rebuilding of "Old Ironsides". (Photo Oregon Collection, University of Oregon)

BRIDAL VEIL FLUME floated sawn lumber down canyon. (Photo Oregon Collection, University of Oregon)

MECHANICAL BUCKER Wade drag saw, one of several gas driven cut-off saws. (Angelus photo Oregon Collection, University of Oregon)

LANGELL'S MECHANICAL FALLER was a noble try but did not replace axe work on undercut. (Photo Oregon Collection, University of Oregon)

IT TOOK TWO TO DRIVE this Winther twin-control log truck on wire-laced plank road. (Photo Oregon Collection, University of Oregon)

"CLANG" GOES GUT HAMMER at Manary Logging Company camp. Hungry crew, washed and brushed, wait for next signal and then fall to. Photographs were taken in a latter day when cook houses had electric lights and cafeteria style service. (Manary Logging Company photos)

ANY GOOD KNEE DIGGERS AROUND?

When sailing ships were still able to pay a profit and even steam vessels were built out of wood, the West Coast supplied quantities of ships timbers and planking. But out of the Willamette Valley and Chehalis River areas came a shipbuilding specialty that has drifted into the past and taken with it the skilled tradesmen who dug up the wood and shaped it for the shipwrights.

Several hundred thousand ships knees were furnished by Oregon and Washington Douglas fir. Shipyards on the East Coast wanted them in hackmatack; on the Gulf and Southwest Coasts it was oak, but fir ships knees were always in demand. There were three types which had been in shipbuilding use for centuries—stem, stern and cabin knees, the latter to join deck beams to framing members.

The trees furnishing these angled buttresses were those which had been prevented from sending down substantial tap roots and instead spread their roots out fingerwise just under the top earth. The knees occurred where the tree trunks bent into the ground.

Knee diggers were skilled at the trade of finding and dressing the knees and the work required clever, accurate axe handling. They dug earth away from the top of the roots and chopped through until the tree fell. They sawed the trunk off and left it. Knees were split out and cut smoothly all around with axe.

The width when dressed determined the knee's value and they ranged from 4 to 24 inches, selling for a dollar per inch. The largest knee on record had 12 feet of stem, 8 foot 8 inches of root, with a 24 inch width. It brought $250 and went into the steamer Ballentyne in 1917.

SHIPLAP FOR THE ORIENT. Freighter West Keats loads lumber at Clark and Wilson mill, Linnton. (Photo Oregon Collection, University of Oregon)

WINDJAMMER AT WARREN. (Photo Weister Co. courtesy Oregon Collection, University of Oregon)

FAMOUS TRIP-GATE FLUME

From 1905 to 1917 the Stanley Smith Lumber Company of Hood River operated what is believed to be the only level log flume using trip gates and sluice flumes to lower logs to intermediate ponds. The entire flume system was about six miles long, built through a heavy stand of principally Douglas fir timber. About 150 million feet of logs were flumed to the Stanley Smith mill at Greenpoint, eleven miles from Hood River at the foot of Mt. Defiance.

Many log flumes have been used in the west to transport logs but they were built on variable grades using swift running water. The Stanley Smith flume had to be built on the level due to the comparatively flat terrain which permitted only a variation of approximately 400 feet for the entire distance. One section of the flume was about 1½ miles in length and the logs were tied together with small rafting dogs and haywire, each turn having from 20 to 40 thousand feet of logs. At the head end of the string, a man with an improvised peavy, helped a small top current to move the logs to the next trip gate.

The lumber from the Greenpoint mill was transported to the Union Pacific Railroad at Ruthton, a short distance west of Hood River, in a 10 mile gravity flume from a 3,000 foot elevation to practically sea level. The storage lakes and ponds of the system are now part of a co-operative irrigation system.

In the Stanley Smith flume, it was possible to get a special log out of the woods, 16 miles from the railroad, and have the lumber product on a rail car within 48 hours. The only transportation facility involved in getting logs from a 3,400 foot elevation to sea level was a horse team and water.

A member of the Stanley Smith firm was Anton A. "Tony" Lausmann who later started the Mitchell Point Lumber Company and with the Miller family of Portland organized the Eastside Logging Company, Rock Creek Logging Company and others in Oregon. Still later he built and operated the Kogap Lumber Industries (Keep Oregon Green and Productive). In jest he says he "worked his way through the woods with a concertina to keep from thinking about all his mistakes." Preserving photographs of early logging operations was not one of them. A few are used in this book.

150 MILLION FEET TO GREENPOINT MILL. Flume and sluice gate of historic Stanley Smith Lumber Company operation in Hood River County. Picture shows logs with rafting dogs removed bunched up to sluice gate. (Photo courtesy A. A. Lausmann, Kogap Lumber Industries)

BIG STANLEY SMITH TRESTLE. Largest trestle in system showing trip gate which controlled sluice into Pond No. 2 as shown on opposite page, top right. (Below left) Pond No. 1, logs being pulled uphill by wide-faced donkey on fore-and-aft road and dumped into pond. Man in foreground is dogging string of man-towed logs. (Below right) After sluice gate was opened, strong current moved logs into sluice flumes and mill pond. (Photos courtesy A. A. Lausmann, Kogap Lumber Industries)

POND END OF SLUICE FLUME (Above left) where logs entered a lower-elevation flume which finally delivered them into Pond No. 1—opposite page, lower left. (Above right) Pond No. 2 where wide-faced Willamette road donkey hauled logs over mile long fore-and-aft road. Smoke from yarder is visible in distance—horse logging landing between donkey and end of flume. (Below) End of flume construction in 1911—lake water feeding flume. Figure at right on old skid road is A. A. "Tony" Lausmann. (Photos courtesy A. A. Lausmann, Kogap Lumber Industries)

THE DEATH OF ROUGH HOUSE PETE

From "Rhymes of a Lumberjack"

By ROBERT E. SWANSON

At the top of a darksome stairway, on a two-bit flophouse cot,
Old Rough House Pete lay down his weary head;
For his rough house days were over, and his last hangup was fought.
Ere he closed his bloodshot eyes, old Rough House said:

"In the days of bull-team logging, when they hauled logs on the skid,
And they cut the stumps away up above the swell;
It was then I hit the jungles — I was only just a kid,
But the roughest, toughest kid this side of Hell.
I could lick a cougar cat, if a man should drop the hat,
And I'd kick my way from jail with flying feet;
California to Alaska — I was famous for just that,
And so, far and wide they hailed me, 'Rough House Pete.'

"When the bull teams all were finished and the ground-lead donkey came,
And the line-horse days had gone beyond recall;
I had hired to do the hookin' — up at Simoom Sound of fame,
But the camp and cook were haywire — grub and all.
On the table-top jumped I, and made the dishes fly,
You can bet my old caulk-shoes weren't very slow.
Then the Cassiar I boarded, and I bade that camp good-bye,
And they stowed me with the cattle down below.

(Continued on Page 137)

OREGON-AMERICAN LUMBER COMPANY (Below and opposite) Plant and company buildings at Vernonia in 1923. (Photo C. Kinsey courtesy A. A. Lausmann, Kogap Lumber Industries)

NO TEN TON TITAN BUT—Muck Riggs, one of A. A. Lausmann's supervisory personnel in Medford, helped to improve log trucking with home made rig in 1927— pneumatic tires used in place of hard rubber. (Photo courtesy A. A. Lausmann, Kogap Lumber Industries)

TREES GO TO SEA. Two steam tugs prepare to tow cigar-shaped Benson raft down Columbia river on its sea voyage to San Diego. (Photo Oregon Collection, University of Oregon)

FLOATING FORESTS

For about 40 years the biggest things floating in the Pacific Ocean were the Benson and Davis log rafts—differing types but each successful. Both rafts served the same ends, to get big quantities of logs from areas where they were logged to sawmills in other areas, quickly and cheaply. The Benson and Davis rafts were born and died in the era of big log production.

The first ocean-going log raft was conceived and built by Captain H. R. Robertson, of St. Johns, New Brunswick. This raft was built in the late eighties in a land cradle patented by Captain Robertson and contained about 110,000 lineal feet of piling, destined for Boston. This raft, although small compared with the later-day Benson raft, was so heavy that a tedious and troublesome time was experienced in launching it.

In 1906 S. Benson and O. J. Evenson, who were principals in Benson Timber Co., entered the ocean rafting business at the mouth of Wallace Slough, Oregon, on the Columbia River, engaging John A. Fastabend to superintend construction of cradle and raft. Fastabend and Evenson constructed a simple cradle with an improved center locking device, and improved upon the towing gear and system used by Captain Robertson. The first raft contained not only piling and sawlogs of all lengths and sizes, but several hundred thousand feet of sawn timbers and lumber for a complete Benson sawmill to be erected

INSIDE EMPTY CRADLE 800 feet long, 30 feet high, 40 feet wide. After being filled with logs, cradle was uncoupled and two halves pulled away from logs. (Photo Oregon Collection, University of Oregon)

TREE-LENGTH LOGS PACKED TIGHT. Derrick at far end lifted logs from Columbia River slough and men inside cant-hooked them into position, threading them with chains. (Photo University of Washington)

at San Diego, California. Evenson accompanied the tug towing the raft to watch its action in the sea, which resulted in several improvements. The first raft arrived in perfect condition and construction started on the mill as soon as the raft could be opened up.

The rafts themselves were 1,000 foot, cigar-shaped structures, expertly engineered. In the quiet waters of Wallace Slough fresh water was necessary as marine borers of salt water would have eaten up the cradle in one season— Fastabend and Evenson built hundreds of these marvels.

John Fastabend had come to Astoria, Oregon, in 1892 from Salt Lake City to take charge of the bridge and building department of a proposed railroad between Astoria and Hillsboro. He did some construction work around Smith's Point but no further progress on the railroad was ever made. When the Benson Timber Co. conceived the raft idea, Fastabend was put in charge of the building operations.

At the cradle grounds in the slough, tree length logs of all sizes were boomed close to a floating derrick and a foreman directed the choosing of the logs. Great skill was required in selecting the lengths and weight of logs and weaving them together to withstand the constant action of ocean waves and ground swells.

The Benson raft was constructed by use of a floating cradle, or form, built in sections so it could be removed from one side of the raft when completed. One side of the cradle was moored to the piling to hold it while the raft was being pulled out. Tree-length logs constituted the greatest amount of the material in the raft, but logs ranged from 20 feet to 150 feet in length, most commonly between 50 and 150 feet long. It required great lift to raise a long tree from the water by use of grabs and the 103-foot boom on the derrick. Sometimes the floating pontoon on which the derrick was stationed was almost tipped over by the weight of the tree.

BENSON RAFT A-BUILDING. Airview of lower Columbia River slough showing completed raft and derrick laying tree-length logs in cradle for second raft. (Photo University of Washington)

Long logs were preferable because they provided a necessary lap and backbone which resisted the action of the water as the logs were loaded into the cradle. When the raft was about half completed a two and one-half-inch stud anchor chain was run through the center of the raft from end to end. Other chains were shackled to the center chain and attached to the circle chains at the end of the raft. A tow chain was attached to one of the circle chains near the end of the raft, providing an emergency tow chain if necessary. The raft towed equally well from either end. Total weight of chain in one of these rafts was about 175 tons; the raft dimensions—55 feet wide, maximum depth 35 feet and approximately 1,000 feet in length. The cradle was 960 feet long, but the peaks of the rafts was constructed with about 30 feet of the logs extending at either end. It required about 30 feet of water to successfully launch one of the rafts from the cradle. Sometimes several of the rafts were constructed during the winter and spring and summer, awaiting arrival of the most favorable towing season, which was June 15 to September 15. Most Benson rafts carried deck loads of cedar poles, spars, shingles, lath ,and occasionally machinery.

READY FOR JOURNEY with heavy chains coupled 800 foot Benson raft awaits river tugs. Note steel nose-plate and chain for towing hawser.

READY FOR DECKLOAD. River boat at left carries machinery which will go piggy-back on Benson raft on Pacific Ocean run to San Diego. (Photos Oregon Collection, University of Oregon)

MYSTERY OF THE MISSING RAFT

One day in 1900, in the San Francisco office of the Pope and Talbot interests, as the story goes, Capt. Hugh Robertson of the Robertson Rafting Co. looked hopelessly at his assistant John M. Ayres and said: "If we can't come up with any better story than we have, we'd better go ashore and buy a chicken ranch."

The fact was they had lost a raft containing five schooner loads of Douglas fir logs. On board the tug Czarina they had towed the raft from Seattle around Cape Flattery and worked south down the Oregon and California coast. Capt. Robertson had the day watch, Mark Schweger, assistant foreman, the night watch. The latter had gone below for coffee. Meanwhile the helmsman had seen flashes of the Cape Blanco light and later watched for St. George's Reef and Trinidad lights. In the early dawn Schweger found the manila towing line cut, the tow long gone.

For two days the Czarina searched the area for the 800 foot raft which was made up of 5 million board feet of fir, and had a beam of 50 feet. Then Capt. Robertson went ashore at Eureka and then to San Francisco. Two other tugs searched the coast line and far out to sea. There was no sign of the raft.

The Robertson Rafting Co. had thousands of dollars at stake in the raft and the search went on, off shore and on. Word sifted through to John Ayres that a sailor drinking down on the Embarcadero was telling a strange story of how he had been paid $100 to stay off his ship.

Ayres hurried down and got the man's story for more drinks. The sailor had been fireman on the steam schooner San Pedro which was towing a big raft of logs when she ran out of coal. Forsaking her tow, the schooner had maneuvered into port, reloaded fuel and meanwhile the fireman started drinking. When he tried to get aboard the schooner, the captain sent him back to shore with a hundred dollar bill and orders to make himself scarce.

John Ayres chartered a tugboat and with the half drunk fireman as a guide, got on the trail of the San Pedro. The next day they came up on the raft which was tight and fully intact. Eventually they overhauled the schooner whose captain agreed readily enough to pull into port, but insisted he found the raft adrift and would claim salvage.

The case dragged on and the log raft remained tied up for many weeks. No evidence could ever be produced against the master of the Son Pedro as having cut the tow line nor could any logical explanation be produced as to how it was cut. The San Pedro was paid moderate salvage money but the mystery remained unsolved.

*　　*　　*　　*　　*

On a stormy December day the big tug Ranger with raft in tow, ran into trouble off the California coast and to keep the logs off the bar at low tide, the tug went inshore. At midnight, trying to make fast on shore, she broke her rudder. After emergency repairs, the ranger went out again and this time was battered ashore. Refloated, she allowed the raft to go aground in the breakers on South Spit. The Ranger, with 600 feet of hawser, kept hold of the raft for several hours and then floated off.

But the trouble was not over. The Ranger could not hold her tow in the wild sea and was forced to cut loose. The raft piled up on shore but the chains held the logs together. Three days later a life saving crew managed to pull the raft away from the jetty but now another hurricane was raging. The tug ran inshore to serve as an anchor and keep the raft out of the billows. The next morning the tow started once more but the raft grounded on the middle quicksands. Two tugs worked at it for five days and finally the tow floated free.

DAVIS RAFTS

Built on entirely different principles, Davis rafts were designed by a Canadian, G. G. Davis, and first used in the rough waters of Queen Charlottes and off Vancouver Island for moving logs from shallow harbors to sawmills. F. A. Douty, president of Multnomah Box Co. in Portland, came across their use and purchased the patents for Oregon and Washington waters. Production pressure of World War I put the Davis method to full test and in 1922 the box concern gained control of patents for use in United States and Canada.

The Davis raft used no cradle and logs were not confined to tree length. The first layer was called the "mat" with "side sticks" holding inside logs in place. These were all laced together, figure 8 style, with wire rope up to 1 1/2 inches in diameter by donkey engine. Succeeding layers of logs were "parbuckled" or lifted by donkey loader into place, each layer two logs less than the one below. Rafts were seven or eight layers thick, the top five or six logs wide. Towed as easily as a ship, they survived all seas.

Davis rafts averaged 600 logs, about 200 feet long, 80 feet wide, the width greater underwater as mat was forced down by weight of top logs. Each raft contained 5 to 6 million feet and used about a mile of line per million.

BUILDING DAVIS RAFTS of Sitka spruce in Queen Charlotte Islands, B. C., 1940 (opposite page). Top photograph shows long "side stick". In both views logs are being "parbuckled" on load. (Photos British Columbia Forest Service)

RAFTING MILLION FEET A DAY (Above) in 1925 to 1927 at Bloedel-Donovan booming grounds at Sekiu. Rafts averaged 5 million feet. Figure on left is Elmer Critchfield who became Davis raft authority after years with Bloedel-Donovan, Merrill and Ring, Crown-Zellerbach. During World War II he was in charge of this work for Alaska Spruce Log Production out of Ketchikan. (Darius Kinsey photo courtesy Critchfield Logging Company, Inc.) Below, lifting Davis raft logs at booming grounds on B. C. coast. (Photo Leonard Frank Collection, Vancouver, B. C.)

(Continued from Page 128)

"Oh, those trips aboard that steamboat, up and down
    this rugged coast;
    How the skipper, mate and deck-hands feared
      my tread.
All the doors were kicked to splinters — each railing,
    stair and post;
    When I'd step aboard each time, the skipper said:
'Man the pumps and test the hose — all the first-class
    quarters close;
    For Rough House Pete has stepped aboard my ship.
I would rather face a blizzard when a North-East
    howler blows,
    Than to have that man aboard for just one trip!'

"Came the days of high-lead logging and the fast ma-
    chines of steam.
    You can bet old Rough House Pete was with the
      first
To be foreman of an outfit, way out near Aberdeen;
    But I met a girl, and then my life seemed cursed.
O, that throbbing urge of Nature — how a man can be
    a fool —
    Can be weak as rotten straw-line in her grip;
For the lion may be monarch, but the lioness more
    cruel,
    With her silken claws, defiles his championship.

"O, I know I went soft-hearted and I tried my ways to
    mend,
    And we lived like honest people on the square.
But a match not sealed in Heaven is most surely doomed
    to end,
    And I came home late one night — she wasn't there.
So I packed my bags; Hell bent to Seattle town I went,
    When I found she'd left with Johnnie-on-the-Spot.
'By the Gods!' I swore, 'I'll kill him!' — Many days and
    nights I spent
    In Seattle's tenderloin, but found them not.

"But the gears of Time roll surely, 'til at last all cogs
    must mesh,
    And Coincidence seems but the hand of Fate;
For in Barney's bar one evening stood that traitor in
    the flesh;
    At the sight of him, my blood boiled hot with hate.
Not with guns we fought that meet, but with fists and
    caulk-shod feet;
    With the joint near wrecked, at last I had him
      down.
Then they drank a toast on Barney: 'To the victor,
    Rough House Pete.'
    Then I took the Cassiar and sailed from town.

EAST SIDE LOGGING LOCOMOTIVE and crew including Sonny McEnery, super-intendent. Camp was at Keasy out of Vernonia on Rock Creek. Entire operation and timber destroyed in Wolf Creek fire simultaneous with Tillamook fire, in 1933. Both fires converged in Salmonberry River area in Tillamook County. (Photo C. Kinsey courtesy A. A. Lausmann, Kogap Lumber Industries)

"Oh those years of fights and boozing, and of all-night
  poker games,
  How I've worked and toiled in sunshine, rain
  and sleet.
How I've blown my toil-earned dollars on a lot of skid-
  road dames,
  And I'm dying now with caulk shoes on my feet.
Yet, all might have been serene, if that dame from
  Aberdeen
  Had led me down the straight and narrow way;
All my toil and tribulation for a family might have been;
  But I'm dying now — and don't know how to pray.

" . . . Queer, how the lights glow dimmer . . . my
  feet feel cold and numb,
  And before my eyes I see my wasted years . . .
I can hear a donkey's whistle . . . Slack her down!
  Hey, roll that drum! . . .
  Wake up there, punk! . . . or else clean out
  those ears.
. . . I'll be late tonight, my darling; for the crew are
  fighting fire . . .

Yes, I'll kill the man who stole my love away . . .
Now my eyes are closing heavy — just to sleep is my
  desire . . .
  Oh God . . . I've tried, but I don't know how
  to pray . . . "

"It was thus this rough-neck woodsman told his tale
  of grief and woe,
  Ere his voice trailed off within that dingy room.
There were none to stand there weeping on the day
  we laid him low,
  And no shaft of marble raised to mark his tomb.
Yet, it's strange how Mother Nature, with her all-
  consuming scheme,
  To her breast will always claim the dust she gave.
For today, a sapling fir tree reaches up its branches
  green,
  While it sinks its roots deep down in Oleson's
  grave."

LOADING LONG LOGS at Rock Creek Logging Company, a subsidiary of East Side
Logging Company at Keasy. Operation used 80 sets disconnected PC&F trucks.
(Photos courtesy A. A. Lausmann, Kogap Lumber Industries)

STEAM and TRACKS

take over the woods

From the wheat stubble and shocks of corn came the roaring, fuming monsters to challenge animal power in the woods. It took a long while and a lot of engineering but eventually the loggers fright changed to full acceptance of the yarding tractor. Before this happened, however, steam had to undergo many phases and eventually give way to gasoline and diesel oil before animals drew their last breath in the woods. The Holt Manufacturing Company sounded a fateful note in 1915:

> "The horse is sliding off the map; his friends
> at last admit it.
> He'll hang around a while mayhap, but soon
> he'll have to quit it.
> For things propelled with gasoline increase
> each day in numbers,
> And Dobbin leaves this earthly scene for
> his eternal slumbers."

BEST STEAM TRACTOR OF 1894. Adapted from farm use to haul pine at McCloud River Lumber Co. Harry Benton operated this type engine as a 16-year-old in yellow pine on southern slopes of Mt. Shasta, was nicknamed a "side hill mechanic". (Photo F. Hal Higgins Collection)

REMINGTON STEAM TRACTOR yarding near Snohomish, Washington, in 1891. Built by D. L. Remington and sold to George T. Cline who used it on a contract for Port Blakely Mill Co. "While it worked satisfactorily," recalls Mrs. Cline, "Mr. Remington did not build another since his health failed. However, his father, Marquis de Lafayette Remington, had built a tractor at Woodburn, Oregon, in 1885, called 'Rough and Ready', successful enough to cause Daniel Best to build on his patents for several years." (Photo F. Hal Higgins Collection)

STEAM TRACTOR WITH MIXED LOAD in Washington. (Photo Darius Kinsey Collection from Jesse E. Ebert)

WESTINGHOUSE UPRIGHT STEAM ENGINE used for skidding, cross-hauling and hauling in fir near Grants Pass for Woodcock Brothers sawmill. Engine had V-belt drive with geared differential, 128 to 1 ratio. Now in Collier State Park Logging Museum. (Photo Woodcock Collection from Collier State Park Logging Museum)

BEST STEAM TRACTOR used in sugar pine on southern slopes of Mt. Shasta around 1900. Harry Benton operated similar engine as 16-year-old, was called "side hill mechanic" as he worked on side hill without shop. (Photo Benton Collection, Collier State Park Logging Museum)

ROBERTS & DOAN STEAM WAGON hauling green lumber from mill to railroad in Verdi, Nevada, 1889. Union Iron Works, Sacramento, built four or five of these engines for Capt. John H. Roberts' River Lines, eventually "purchased" out of business by Southern Pacific. Engine had tremendous power, once out-pulling a locomotive. (Photo Mrs. Lowell from F. Hal Higgins Collection)

55 HOLT STEAMER hauling from mill to railroad on construction job by Holt Freighting Co. Front trailer was equipped with auxiliary engine adding 25 to 30% more power for steep grades. Holt also developed a steam braking system for heavy down grades. (Photo Caterpillar Tractor Co. from F. Hal Higgins Collection)

In 1893 the C. L. Best Tractor Company put its first tractor into the logging woods and by the end of 1894, Holt and Best had four machines operating in Oregon and California. These were the big-wheel behemoths that pulled wagons and had to operate on good, level roads. About their only efficiency was they cost less than railroads.

But before the Best and Holt era, there were several steam traction engines that made history in the timber if not immortality. One was the English Aveling and Porter. George Gable of Oroville, California, owned one which was used by the Merrimack-Pea Vine Lumber Company, according to the Oroville Public Library. This machine cost $5000, was shipped "around the Horn", was naturally nicknamed "Johnny Bull" and its use in the woods continued to about 1890.

Records of F. Hal Higgins, collector of tractoriana, show a steam traction engine built by Owens, Lane & Dyer on exhibit in a private museum at Angels Camp, California. Nicknamed "Jenny", the machine started out in mining at Placerville and in 1896 passed into the hands of N. A. and J. E. McKay, sawmill men, who converted it to logging. Legend has it that the bull team drivers cursed it harder than they did their own bulls and one of them was killed trying to outlog it his own way.

While not strictly logging machines, the Roberts and Doan's Steam Wagons had colorful history as freighters which defied the railroads. Doan had brought a model steam traction engine to Sacramento, where he worked for the Union Iron Works. He persuaded Capt. John H. Roberts, a river boat freighter to "build my steam wagon and whip the railroads that are stealing your business." Roberts did and in 1893 had perfected the 3-wheel affair to the point where five of them were freighting goods on the Sacramento Valley roads as far north as Chico to feed Roberts' river boats.

"The late Mrs. Lowell, daughter of Doan," says Hal Higgins, "maintained that the Southern Pacific Railroad paid Roberts $20,000 to abandon his steam wagon system in the Sacramento Valley. This seems logical since Roberts took them across the Sierras at Emigrant Gap and then on to Verdi, Nevada, a few miles this side of Reno to feed the railroad by hauling lumber from mills to railroad siding. The power of these all-wheel-drive Roberts & Doan steam tractors was something to talk about for years after they disappeared. On level ground, they hauled as high as 12 heavy loads of green lumber over plains and mountains, 16 miles twice a day.

"On the steepest grades they unhooked and cut the load to four wagons. Their loud noise and flickering headlights made them a fabulous sight and sound. Artists from Harpers and other magazines came out from New York to sketch them. But to the blase railroad men who operated the glamorous shiny big locomotives on rails, the big 3-wheel monsters were a joke. To prove it, a tail-to-tail tug-of-war was put on. When the big rough-neck land steam wagon pulled the rail locomotive backwards as far as it wanted to go, there were no more funny jokes from the SP engineers. Charles Root, who was a member of Root, Neilsen owning the Union Iron Works, recalls that the last Doan wagon built cost Capt. Roberts $34,000. These rugged pioneers worked and made history at freighting grain, wood and lumber for at least 20 years."

Out of San Francisco about 1903 came a tractor that went into the woods of the upper Sacramento valley to make some progress for several

ONE OF FIRST TRACK-TYPE TRACTORS—Holt No. 1522 owned by M. G. Ashley.
(Photo Caterpillar Tractor Co. from F. Hal Higgins Collection)

CLOSE VIEW OF EARLY HOLT as shown at work in above photograph. (Photo from Oregon Collection, University of Oregon)

years. This was a steamer something on the style of the 3-wheel steam Best that was beginning to take hold in the woods and mines. It was the Mc-Laughlin and had a front wheel attachment to the frame that was enough different to rate a patent. Recollections of old time tractor men are that it was so near a copy of the Best that the latter threatened infringement proceedings and it faded out. It had a few logging jobs up around Oroville and McCloud, and one is reported to be standing in the edge of the woods on the road east of McCloud.

Various Case, Buffalo Pitts, Advance-Rumely, and other steam thresher engines nibbled at logging in some of the border areas where farming and logging light timber made their use practical for the small operator.

In 1904, the first track-type machine was made at the Holt factory in Stockton—a 40 h.p. steam unit with driving tracks at the rear, front wheels to guide it. It pulled a third more than its 60 h.p. all wheel predecessor. Many of those units went into the woods to haul wagons. In 1905, Holt manufactured the first gasoline track-type tractor and the name "caterpillar" in this connection was born.

In 1908, a logger operating in a scattered stand of timber in Oregon, decided to try the machine for skidding logs to the landing, loading them on wagons and hauling them to the mill. This proved satisfactory, considering the bulkiness of the tractor at that time, but with the large tiller wheel stand-

TRAILING LOGS AT MENDO-CINO Lumber Company operation, Mendocino, California. (Photo Caterpillar Tractor Co. from F. Hal Higgins Collection)

WHEEL OF HALF STEAM TRACTOR. Copy of photograph in Pliny E. Holt's album recovered from Stockton city dump. (Photo Caterpillar Tractor Co. from F. Hal Higgins Collection)

ing out in front it was hard to move the machine for ground skidding amongst the stumps.

The first track-type tractor without the tiller was designed in 1911 and this new development opened up a much larger field for tractors in the woods, as it permitted them not only to haul the logs on wagons, trail them in chutes and to groundskid the logs directly from the stump. Lumber companies in Idaho, Montana and Oregon started immediately to use tractors to replace animals in the woods entirely. One operator in Washington stated in 1913 that there was not a horse on his job.

The idea of ground skidding logs quickly brought up the point—if a small wagon could be used behind the tractor, it would be possible to go into the woods and load the logs on wagons and haul them directly to the mill or the railroad. The first wheeled unit built for this class of work was a "spool wheel bummer." The system used then was to have the tractor pull the bummer to the log, cross-haul it onto the bunks, and then pull the log to the railroad. This proved very satisfactory, as it got the biggest part of the log off the ground, and made it possible to haul heavy loads much easier with the machine. Some loggers used drays and sleds to carry the front end of logs.

Up to this time, however, tractors were practically used only as horse logging units had been. The question developed then as to the possibility of designing a track-type bummer to be used with the "caterpillar". Accordingly, The Holt Company built such a wagon. It was first used in the Potlatch Forests in Idaho. Much larger loads were hauled and the tractor could

CAT LOADING RIG. Early attempt to use Holt gas tractor as loader with gantry carried on front. (Photo F. Hal Higgins Collection)

be used to much better advantage in working on soft ground. The small round wheel bummers were quickly discarded for the track-type.

The fever by this time had gripped the whole logging industry, and the success of tractors in the woods was definitely recognized. On many logging operations, however, they were using slip tongue wheels with horses. The question naturally arose in northern California—why not use tractors instead of horses with these wheels. A machine was connected up to a set, and the increased output was so astonishing that several tractors were sold in California to replace horses on this type of work.

In 1923, with several hundred tractors logging in the United States, Sherman Thorpe, a California logger, and Tom Robinson, a "caterpillar" salesman, devised the idea of mounting a hydraulic pump on the rear of the tractor, and with a hose connection running to the arch above the wheels, have a hydraulically operated sheave take the place of the old slip tongue principle of raising and lowering the logs.

The Best Company designed this attachment and it proved a success right from the start, with much larger footages of logs hauled per unit each day. These all-metal, hydraulic wheels were built in two sizes, one to handle long logs, the other short ones. Records show these wheels could straddle bunched logs and pick up and carry as much as eight thousand feet of logs on half mile hauls, as much as fifty and sixty thousand feet handled with one unit, a tractor and hydraulic wheels.

But these wheels were impractical in many places where the terrain was so steep that straight ground skidding was still followed. In Washington the idea was conceived of building a boiler plate about 6′x8′ square, with

EARLY TRACTOR AND 8-WHEEL WAGON at Washington logging camp. (Photo Darius Kinsey Collection from Jesse E. Ebert)

HAULING LOGS ON BUMMER. Best 60 tractor with 5000 feet of logs used by Blair and Anderson, Union, Washington. (Photo Caterpillar Tractor Co. from F. Hal Higgins Collection)

the nose turned up at the end, and to roll the logs on the pan and thus prevent them from digging into the ground. This eventually led to the use of a winch which fitted directly on the rear of the tractor. A cable was put on this winch and whenever it was found that the tractor could not get logs out over a certain pot hole, or off steep hillsides, the timber was line-logged out of these control places.

With this idea, the double drum was developed and made into a versatile skidder for line-logging timber out of bad places. In 1924, Gail Spain, a young mechanical engineer with the Willamette Iron and Steel Works, conceived the idea of manufacturing single and double drum logging winches. These were used on all Best tractors.

Another development came when "Jack" Meister, of the Shevlin-Hixon Company of Bend, Oregon, conceived the idea of mounting a fairlead on top of the wheels instead of a hydraulic lift. Thus, with a winch mounted on the rear of the tractor, he line-logged the timber in and up under the arch. This eliminated the bunching cost. This idea quickly spread, and by 1927 dozens of fairlead arches were in use on the Western Coast. Then as time went along, fairlead skidding bummers were made.

The tractor had now proved itself successful and with other manufacturers in the field, passed into succeeding stages of refinement, taking full control of West Coast logging and witnessing the end of animal power and cumbersome rail equipment.

THE PUSH

By

PAUL HOSMER

Newspaper man turned toward industry, Paul Hosmer has seen pine logging change from the days of horses and high wheels to 230 h.p. cats and all during this time he has produced a little gem of a house organ for Brooks-Scanlon, "Pine Echoes."

Paul is gifted with a needle-punch sense of humor and the ability to make loggers and lumber men something more than story book characters, something like people with human foibles. Many of his contributions to national magazines and sketches in "Pine Echoes," such as the one following are No. 1 select literature.

A sawmill is a place where lumber is cut from logs, contrary to the opinion of a number of uninformed people who still believe that lumber comes from a yard. Some years ago a mill that could cut 10,000 feet of lumber in a day was classed as a leading industry, but today even the ordinary sawmill will turn out from 200,000 to 600,000 feet each twenty-four hours. While a great many people don't realize it, the fact is the job of supplying enough logs to make that much lumber every day is quite a chore, and practically all the waking hours of one man, in between cruising trips, logging congresses, forest fires and other interruptions, is spent in looking after the details on the payroll as the Logging Superintendent. Around the woods he is referred to as the Push.

The logging superintendent is one of those men you read about in books, provided you read those kind of books, and is a master of a hundred and eight different trades and professions. He knows more about each one of them than the man who invented it, and has as much influence around the outfit as a man with a case of Scotch at a session of the State legislature. He has to be able to lower a locomotive down a six percent grade with three cars of logs and no air; he has to be able to pace down two sides of a section and guess how many feet of logs there are in the middle of it, and if he guesses wrong he receives a dirty letter from the owner; he has to be able to design a new and heavier radiator plate for tractors after he finds that the one with which his fleet is equipped will not withstand the shock of being driven into a seven-foot fir stump without bending; he has to be able to advise the proper remedy for the haul-back team the morning after it has chewed the end out of the spare bag of oats; he has to be able to stand in the middle of a tamarack swamp in a state of partial eclipse, with a map in one hand and a compass in the other, and tell where to build a railroad that will stand up under a load of logs without costing the same per mile as the New York Central. He talks in larger figures than the treasurer of General Motors and can divide 875,000,000 feet by $4.35 per-thousand in his head as easily as a garter snake sliding through a puddle of red engine oil. He has paced down so many township lines that he counts automatically as he walks and his shoes go to pieces faster than a mail order tire.

The logging superintendent maintains a number of homes, domiciles and places to sleep. He has a town house where the wife and family live and which he visits occasionally and from time to time. He moves a bunk and a couple of blankets into the rear end of the timekeeper's shack at the headquarters camp and stays there whenever he is too tired to go home, or whenever he learns that the cook is having hotcakes and fresh pork sausage for breakfast. A tent, a cooking outfit, some extra blankets and a pair of snowshoes are kept ready for use in the company warehouse and these things constitute the luxury of his home life while on cruising trips or other punitive expeditions into the brush. At all other times his home is wherever his hat happens to be, which includes a number of most unique places. At various times his hat is liable to be seen bobbing around on the running board of an engine, flapping dolefully from the upper deck of a pine stump as he superintends the replacing of three wrecked logging flats on the track, or waving wildly in the mid-summer breeze as he deftly flags the steel gang speeder on a down grade and bums a ride into camp.

He goes about weighted down with eleven pounds of gadgets, doo-dads and thingumajigs cached around in various pockets in his clothes. Up in Washington the body of a logging superintendent was once found hanging in the lower branches of a Douglas fir, where it had been carelessly flipped by a casual mainline, which yielded some fourteen pounds of odds and ends, amongst which were included a safety razor, three dozen matches, a box compass, one two-bit pipe, a belt axe, seven pieces of colored chalk, a ball of string, a handful of shingle nails, one second-hand cork, a trip block pin, six log scale books, a track bolt nut, and a plug of Climax.

REPLACED SLIP TONGUE. Hydraulic-controlled arch for raising and lowering logs built by Best company which also designed early tractor-drawn trucks. (Photo F. Hal Higgins Collection)

While the logging superintendent is a 14-karat woodsman who has been logging ever since Babylon fell and Tyre was punctured, you will never guess it by looking at him. Camp foremen, time-keepers and other wage slaves maintain the traditions of the old time lumberjack in the way of dress, but the push is just as liable to appear alongside the jammer in a straw hat and a pair of tennis shoes as he is in a white collar and a pair of tin pants. There is nothing proud about him and he borrows chewing tobacco with equal readiness from the bullcook or the owner of the outfit. He is the possessor of the only charge account on the books of the camp commissary, where he gets his personal supplies such as shoes, clothes and gloves. The storekeeper personally makes up all losses due to shrinkage in stocks of peanuts, chocolate bars, hard candy and smoking tobacco.

His manner of speaking is large and indefinite and he talks of geographical points in the same obscure manner as Captain Byrd telling how he flew over the Pole. When the travelling agent for a saw company asks him in a respectful tone how to find the fallers at Camp Six, the logging superin-tendent points haphazardly with his thumb in a general northerly direction and tells him the fallers are working right over there about a mile and if he will just follow that trail a short ways he will hear the trees falling. The chances are the fallers are three miles down the road and a mile and a half off the trail, but distances mean no more to a logging superintendent than fallen arches to an an-gleworm. The loading crew, he informs a cigaret salesman, are working on the spur which runs up into the NE-NE of Section 17, which is really quite exact, provided you are armed with a map and compass and were born in the woods to start with. The logging superintendent is oriented with the world from birth and can be blindfolded and tossed into the middle of a cedar swamp at midnight without disturbing his sense of direction. Looking calmly around him he will take a chew of tobacco, snap his suspenders three times and start off on a bee-line through the woods, and inside of three hours he will appear on the front steps of the cookhouse in time for breakfast. If it is one of the days when they are having hotcakes and fresh pork sausage he can beat that time by twenty minutes.

FIRST BROOKS-SCANLON ARCH. First experimental arch being operated with cat near Bend as Brooks-Scanlon officials check results. (Photo Paul Hosmer)

FORDSON STUMP PULLER.
(Photo Oregon Collection, University of Oregon)

TRACTOR-MOUNTED STUMP SAW.
(Photo Oregon Collection, University of Oregon)

The logging superintendent suffers from two secret sorrows in life, both of which cause him some anxiety and loss of sleep. One of them is the log pond. It is his duty to see that the pond is kept full of logs in order that sawmill may never have to wait, and a certain polite, but exceedingly intense, rivalry is maintained between him and the mill boss. They go through life in a state of armed neutrality. The mill boss only has one big ambition and that is that some day, in some way or other, something will happen to the logging department that will allow him to cut every log in the pond and have something to hold over the logging superintendent the rest of his life. The latter lives in the hope that he can keep a jump ahead of the sawmill and every morning on the way to the woods he drives by the pond to make sure that the mill boss hasn't slipped over an extra shift on him or added a couple of new band saws without telling him about it. If he finds that one of the trains failed to get in, due to a log falling off and jill-poking a couple of cars off the track, the logging superintendent pulls his hat down over his eyes, snaps his suspenders six times in succession, shoves the old car into high and takes out for the woods so fast that he dips sand with the back seat on every curve. Rushing about from camp to camp he urges his men to renewed efforts in the way of logging, with the result that that night sixty cars of logs go into town and fourteen of them have to stand on the sidetrack two days waiting for the pond to open up enough so they can be unloaded. The next morning, as he drives by the pond, the logging superintendent's eye lights up with satisfaction as he notices there is no water showing, so he drives contentedly into camp with nothing on his mind to worry about except the Forest Service, which is the second of his secret sorrows.

From the superintendent's viewpoint it is more or less discouraging to move in on Section 14 with six newly purchased cats only to discover that the Forest Service does not allow caterpillar logging on that section. He gets a slight jump on the situation, however, through the fact that it is three days before the government inspector can order him off, due to the number of reports he has to make out before he can go into action. There is a certain amount of red tape which has to be gone through, and the push manages to load out thirty cars of logs before the inspector gives him the official grand bounce. Then the government man looks over the damage, adds a fine of two-bits a tree, a penalty for mussing up the ground, a tax for cutting government timber without a permit, and multiplies the agreed stumpage price by two, after which he hands the bill to the logging superintendent with a leer. The latter moves his cats over onto Section 22, which he hadn't intended to log until next fall, makes a couple of dirty cracks in an undertone about government men, and goes back to town where he enters into an earnest conference with the Forest Supervisor. The Forest Supervisor listens calmly to the superintendent's arguments and finally remarks that, while he can't do anything about it himself, he will be glad to write a report of the incident to the District Forester and ask for instructions. Ten days later he receives a reply on Form 122-A-393BB, stating that while Section 14-32-19 seems to be free from

BELT-DRIVE WOODSAW mounted on 1908 Oldsmobile chassis. (Photo Oregon Collection, University of Oregon)

blister rust, the Washington office has received a report that Section 15-32-19 is heavily infested with the pine beetle, and will the Supervisor please make a report on this disgraceful state of affairs.

Something appears to be haywire with the works, but on looking up his files the Supervisor finds that Form 122-A-393BB is the wrong report blank, so he sends in another report calling attention to the error. Ten days later he receives blank form No. 1344-BF-1894XX-BVD, calling for a report on the matter of caterpillar logging on Section 14 in the right state, and the affair, having thus got started on the right track, is now officially and legally opened for discussion. Reports, cruising estimates, personal letters and blank forms fly back and forth across the country during the following three months and at last the logging superintendent receives an official notification to the effect that cats will not be allowed on Section 14. Receipt of this ultimatum tickles the superintendent so much that he bursts forth into uncontrollable paroxysms of laughter and hies himself to his bunk in the time-keeper's shack, where he gropes shakily around under his mattress and brings to light a small flask of snake bite medicine. He takes three hefty snorts in rapid succession and slowly regains his composure, after which he sits on the edge of the bed with his head in his hands and offers up a short but fervent prayer for all government officials, from the Chief Forester down to the buck privates.

Another thing that keeps him in a state of nervous tension when the forestry men are around is the fact that the government makes him pile and burn all brush and slashings. As soon as the logs are off the superintendent has to put in a crew of expert gardeners, fancy hairdressers and manicurists to carefully rake the ground, pile the brush neatly and sweep up all debris with a broom and a dustpan, after which a crew of firemen burn the slashings on certain hours, days and weeks of the calendar as designated by the government. Added to the fact that he can only cut certain trees on a government section, the manicuring of the ground runs the superintendent's logging costs up another dollar and a half a thousand, which of course fills his soul with gladness and he is as pleased as a Chicago beer baron who has just been asked by his wife to feel around under the front porch and see what is making the noise like a clock ticking.

One outstanding idiosyncrasy of the logging superintendent is the fact that he insists a cruising trip is fun. Where he ever collected such an idea nobody knows, as every time he returns from one of these delightful excursions into the great outdoors he has to spend three days in bed resting up, and oftentimes it takes the compassman and cook a week to recover. Loaded down with 70 pounds of blankets, canned food, axes, note books, maps, waterbags, tin pails and a frying pan, he paces down section lines by the hour, counting trees, estimating stumpage, and fighting deer flies and mosquitoes with both hands as he scribbles the result of his guesses in a note book. He gets so used to counting paces that when the cook wakes him up the second morning out to ask whether he should make flap-jacks for breakfast or open another can of beans, the only answer he gets is a mumbled "—twenty-two, twenty-three, twenty-four," as the superintendent rolls over and unconsciously slaps at an imaginary horsefly. He thinks nothing of hiking twelve miles through the brush with all his possessions on his back and hung around his person. Two weeks later as the compassman and the cook rub liniment on each other's backs they are unanimous in agreeing that they don't think so much of it either.

There is only one thing about cruising that makes him mad and that is when, in answer to a glowing letter from some enthusiastic owner about a forty which is for sale cheap, he lopes across two pumice flats and climbs three mountains only to discover that, while it appears to be an excellent logging chance if there was any timber, the only thing growing on the place is a cluster of Indian Pink in the center and a family of skunks over on the east line.

For some obscure reason the logging superintendent's pet love is the camp telephone line. This modern convenience consists of a thread of light wire originally hung on saplings stretched between the main office and the camps, but which for the past several years has been squirming gracelessly along the railroad track, waving gently in the snowbrush or wriggling over the rimrock and down logs where it blew off the poles back in 1924. Occasionally a timekeeper, grown desperate over not being able to talk to his girl in town, dashes down the line and ties it together where a careless faller has dropped a tree across it, but it takes some such a crisis as this to make him do it. Nobody has ever yet been able to get the camp they think they are going to and it wouldn't do any good anyway as they couldn't hear anything if they did. The camp telephone is one of those things the logging super-intendent is always going to fix just as soon as he can get around to it, but in ten years he has never been able to remember it.

Only occasionally will the superintendent admit that there is anything the matter with the line, these rare instances usually occurring just after the Forestry Office has called up to ask what all that smoke is up around Camp Six. The superintendent takes down the telephone receiver and rings six times with the bell handle and waits expectantly for an answer. The first thing he hears is a slight buzzing sound something like a horsefly circling the field while looking for a good landing place, but this only lasts a moment. Just as he is straining every nerve to catch the voice of the timekeeper which, in spite of previous experience he firmly expects to hear, another jolt of static knocks his ear drum a quarter of an inch out of line, and reminds him of a stevedore falling down a gangplank with

MAGNIFICENT SPRUCE in Manary Logging Company stand.
(Manary Logging Company Photo)

a load of tin dishpans. This is followed by three loud squeals and a violent pop, which, bursting on his unprotected ear, causes him to see two pinwheels and a couple of Roman candles just as three brilliant red, white and green balls of fire descend from the ether and blow up directly back of his right eyeball.

In the midst of this pandemonium he hears snatches of what might be several excited voices floating in from afar and the logging superintendent shouts mightily into the phone. When the heavy artillery dies down for a moment he identifies one of the chorus as the voice of Yulius Yonson, of Camp Two, who appears to be in distress due to the failure of a rush order of Copenhagen snuff to arrive on time. Another voice he recognizes as belonging to a certain Michael Gallagher, foreman at Camp Four. Michael is complaining, as usual, of the haywire machinery they send him. Still a third voice, calling frantically from Camp Three, proves to be one Tony Sorrentino of the section gang, who appears to be stuck some place with a gas speeder. The logging superintendent summons his self-control and listens while something like this explodes in his sore ear:

"Hey! Aye vant to know for vhy Aye don' get de t'ree box Copenhagen snoos—Hey, wotsa mat'? I pusha da crank, I ringa da bell, I maka do biga longa leesen—an' them damn skiddin' tongs you been sendin' us ain't no good. They keep bustin' on—on de train an' she don' ban dar an' Aye go down to Camp Vun—an' I don' getta somebod'. I can'ta get da Casey-Jones to maka da begin—an' the black-smith says he can't fix 'em no more. How'nhell am I goin' to do any logging—an' Aye see Mister Yones an' he don' have no Copenhagen—hey, wotsa dis, getta my foot offa da track? I no hava my foot on da track—Aye lose two gude men dis morning an' two more is goin' to qvit tonight—with a six cat show an' a haywire outfit—I maka da biga push wid da handle but it no maka da begin—Bay Yeesus, Aye got to have snoos. She ban hal of a vay to run loggin' camp. Aye bat you—Meester, pliz, wot I do, hey?—an' them chokers keep breakin' on a four-log hitch with a five-way butt hook— Bay Yiminy, Aye tal you—"

With a sigh the logging superintendent hangs up the receiver, mumbles something under his breath about having to fix that thing just as soon as he can get around to it, and beats it for camp in the flivver to see about the smoke himself. He can crawl to the fire on his hands and knees with an anchor tied to each foot faster than he can find out anything over the telephone system, but he nevertheless retains a sublime faith in the line and knows that it will work all right if he could just remember to have it repaired.

But in spite of the trials and troubles of his life and the moral and mental turmoil into which he is plunged by glimpses of blue water in the log pond and blank forms from the Forestry Office, the logging superintendent manages to cling to life with an amazing tenacity and continues to dump logs into the pond with wild abandon. Every so often the mill foreman jumps on his neck with blood in his eye and breaks the news that a $300 band saw almost got ruined yesterday through the fact that the starspangled roughnecks up in the woods left a spike in a log. The superintendent bites off another chew of Climax, snaps his suspenders a couple of times for luck, and wanders over to the un-loading dock where the boom man is idly strolling about on the floating logs setting out fish lines. During the next few minutes he painstakingly points out to the boom man the folly of driving ten-inch boom spikes into logs to hold a ten-cent fish line when a shingle nail will do just as well, and any-way, if he has to use a boom spike, will he please remove it before the mill foreman has another hemor-rhage, after which the matter of wrecked band saws is erased from his mind until such time as the mill foreman ambushes him again and shows him a broken choker which the skidder crew was too tired to unravel from a butt log.

He takes up the matter of broken injectors with the master mechanic and together they sit in the shade of a logging flat making little scratches on an old brake shoe while discussing the probabil-ity of ever perfecting the 11x14 Diesel yarder to the point where it will run on axle grease. He makes an investigation into the pig situation at camp and discovers that two of them have the backache or some other household ailment, so he holds a consultation with the bullcook and advises the proper treatment. Then he rambles through the cookhouse in a desultory manner, pausing here and there to talk to a waitress, and emerges through the back door. Behind the root cellar he shakes himself like a cocker spaniel coming out from under a garden hose, and sheds part of a lemon cake, four cookies, two doughnuts and a piece of mince pie which were concealed under his coat. The most important thing, however, gained in his tour through the cookhouse is the information to the effect that tomorrow is the day for hotcakes and fresh pork sausage. As he sits on a pile of ties in back of the commissary munching on a doughnut, he decides that he is too tired to go back to. town that night, thereby insuring for himself a ringside seat at the breakfast table.

While the daily life of the logging superintendent is not a tranquil one, it does have a com-pensating feature or two. When everything goes wrong and he gets so downhearted he can't even talk to himself without quarreling he can always sit down at his telephone and try to call up some-body and he can always horn in on hotcakes and fresh pork sausage about three times a week, which is something, you will admit. Even a logging superintendent has his moments.

CALIFORNIA

FROM CHOPPERS TO DOG-HOLES

By 1875, the big redwoods were being logged by a race of men in a forest world of their own. The first man to see possibilities in the immense stands of redwood was probably John Dawson, who deserted ship in 1830, married a Spanish girl and went to whipsawing logs near Bodega.

Most of the early loggers were bachelors who felled a few trees and got enough rough-hewn lumber to build themselves ten by twelve shanties, perhaps a barn and slabs to fence two acres. They worked twelve hours or more a day and dropping an eight foot redwood was a good day's work for a man of even that hardy breed.

The tools they used were simple enough. The entire logging equipment for the night and day run of the mill consisted of two teams of bulls—each team five yoke—with the necessary chains and jackscrews, a few snatch-blocks, some manila rope, and of course, axes and crosscut saws.

In his memoirs, C. R. Johnson, founder of the Union Lumber Co., gives a first hand description of the skidroad work.

"The jackscrew was an important tool in the woods. It was wonderful what two men, each with a jackscrew, could do to a log. They could get it out

NEWPORT CHUTE IN THE '70s at Stewart, Hunter and Johnson's mill on the Pacific cliffs. Schooners were moored by several lines with slack enough to allow running twenty or twenty-five feet with the sea. At lower end of chute was "clapper man" who operated a brake-like device to slow up and stop each stick. Schooner of this size could load 75,000 to 150,000 feet. (Photo Union Lumber Company Collection)

FIRST SAWMILL AT FORT BRAGG. Constructed in 1885 by Fort Bragg Redwood Co. Mill burned in 1888, was replaced and company merged with Noyo Lumber Co. in 1891 to form Union Lumber Co. (Photo Union Lumber Company Collection)

from a hole and turn it clear around. Those jacks were also used where the ground was very steep in jackscrewing down to where the bulls could get the logs. A good jackscrew, as I remember it, cost about $75.

"The cattle also were wonderful. The bull puncher was the highest paid man the country had. A good driver got equal pay to the foreman. When he wanted his team to start a load, he would commence his antics; yelling at the bulls, and jumping up and down and hitting them with his goad stick until finally getting them all started and pulling together. And it was a pretty good load that they wouldn't start.

"Depending upon the size of the logs, a load would consist of from four to eight logs. Each log was sniped a little on the forward end, the head log being the largest, and decreasing in size as they went backward. The logs were fastened together with a short manila line having a 'dog' attached to each end. The dog was driven in through the logs.

"The sugler was another important man. He accompanied the load down to the landing, and his job was to throw water ahead of the load. For this purpose he had a long stick which he carried over his shoulder and a bucket attached to each end. From these buckets he threw water on the road just ahead of the load. Water barrels were located at convenient places so he could often replenish his supply. Where the road was very steep, not much water was needed. But in the comparatively level places, it was a great help—to make the logs slide over the wet places. Where the road was nearly level, skids were put in—sometimes running lengthwise with the road, sometimes across the road. Also, chains were attached by 'dogs' to the log, and on steep places

McGIFFERT MOBILE LOADER AT WEED — 1907. Machine rested on 4 shoes which were set down on ties. Wheels were lifted and empty cars passed underneath loader. Operator had to be "catty" to drop log on swing at proper time on far side of car. Loaded all from one side, averaging 100,000 feet a day. Loader had crew of 6— two hookers, top loader, engineer, fireman and toggle knocker. (Photo Parsons Collection, Collier State Park Logging Museum)

were dropped, to act as brakes and prevent the logs from piling onto the bulls. The suglers were very expert at their work. They had to be nimble-footed, and quick in replenishing the water in their buckets and throwing it under the logs."

Many lumber companies with mills located on streams followed the practice of rolling their logs into the almost dry stream beds in the summer and fall. Then when the winter rains came, the streams flooded and the logs were carried down to the mills. It was necessary to construct splash dams with booms to hold the logs and prevent their going out to sea. Occasionally strong floods would break a boom and all a mill's logs would be lost.

One of the reasons for cutting high stumps in the early days was to eliminate the heavy butt logs, which sank to the bottom and clogged up the streams and log ponds. (After 75 years many of these sinkers were dug out

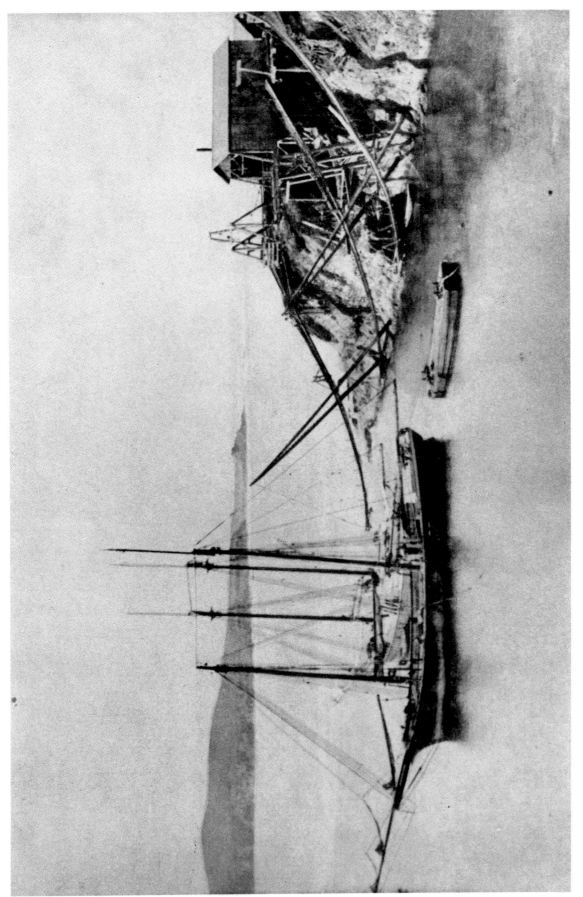

SCHOONERS LOADING FROM APRON CHUTES, MENDOCINO HARBOR—1865.
(Photo Union Lumber Company Collection)

BLOCK AND FALLS NOYO POINT CHUTE which controlled lumber let down wire chute to ship. (Photo Union Lumber Company Collection)

of the mud and manufactured into grape stakes and pickets). Another was the extreme flare at the base of many trees, which made sawing difficult. Also many old trees had hollows (goose-pens) which made attempts to cut low stumps dangerous for the "choppers," as the redwood fallers were known.

About 1875 the first logging railroads were built: on the Noyo and from Caspar to Jug Handle Creek for hauling logs to the mills; on Salmon Creek and at Cuffy's Cove for hauling lumber to the coast. Within a few years almost every major mill had its logging railroad and the number of oxen in use rapidly dwindled. The early railroads were often close to the streams, and were frequently damaged by high water.

Getting the logs into lumber was a challenge to power saws and pioneer engineers. The earliest mills had muley saws, sash saws, or both. The muley saw was a long straight saw pulled back and forth from one end by a water wheel or steam power. Although slow by modern standards, it was a vast improvement over the whip-saw used by hand. Sash saws were either single or in gangs held in a frame resembling a large window sash.

17½ FOOT REDWOOD FELLED IN 1933 below Shake City on Noyo River. Head chopper Emil Johnson (extreme right), second chopper Mike Mantilla (center with axe), chopping boss Henry Gordon (on top of tree). Bill Touchel (extreme left), R. Charlie (left of saw), Charlie Hanula (right of saw), Swanie Martin (bottom). (Photo Union Lumber Company)

MAIN DAM ON HEADWATERS OF BIG RIVER IN 1909. Large bull wheel on right used to raise dam gate. A system of 26 dams was employed on Big River watershed to float logs downstream to a boom where rafts were made up and towed the balance of the distance to the mill. Dams on tributaries were tripped in time to allow the water to hit the main river in unison. (Photo Union Lumber Company Collection)

To get the lumber to market it had to be loaded on ships. Fire chutes usually were used to load lumber, carrying it from the top of one of the bluffs characteristic of the Mendocino coast to a ship anchored in a small cove below. At Newport, lumber was hauled from the Stewart, Hunter and Johnson mill by six-horse teams driven with a single line, called a "jerk-line"—from the mill to the landing, where it was loaded by a gravity chute extending from the shore to the ship below. There was no wharf, the vessel moored fast by lines. There was enough slack in these lines to give the schooner a chance of running back and forth twenty or twenty-five feet with the waves. The apron of the chute projected slightly over the deck of the schooner, and the lumber was sent down a distance of eighty feet, one piece at a time. At the lower end of the chute was a man called the "clapperman," who operated a brake-like device which slowed up and finally stopped each piece of lumber just as it reached the apron. Here the crew of the vessel would take the lumber and stow it.

These coves and inlets were known as "dog-holes" in the Mendocino coastal traffic, and many an early ship was wrecked trying to get in or out of them in stormy weather. A few localities had wharves, for example: Navarro River, Little River, and Usal Creek, but these were in the minority.

FLAT OF DOWN REDWOODS. Note man in lower right hand corner. (Photo Union Lumber Company Collection)

The masters of the early sailing vessels were a colorful and hardy lot, often referred to as commanding California's Scandinavian navy. Long famous was Capt. "Midnight" Olson, who could drop his ship into the worst doghole on the coast in the black of night.

The schooners took all manner of risks in the dog-holes and loading along unidentified cliffs. Many times, with storms making up, they would have to stop loading abruptly and escape to the open sea. Many ships would have to lay off a harbor for days waiting for the seas to calm.

Sometimes an indentation had to be blasted out and ships tried to load with the breakers roaring and pounding into the caverns, the wind whistling and shrieking like devils in the rigging. Many times a schooner stood off shore with a cargo of redwood as a gale blew up. The anchor would break loose, crew hoisting sail but the ship would be a wild sea horse, the masts snapping, the waves flinging her against the cliffs, white crested swells spilling her crew and cargo into the fury of the caves.

NEWELL'S TEAM ON REDWOOD SKID ROAD. (Photo Newbury Mill Co., courtesy Pacific Lumber **Company)**

By 1876, 1,106 vessels had crossed the Humboldt Bar. In 1867 alone approximately 46 million board feet of lumber were shipped from Mendocino coastal ports, largely to San Francisco. Here it received a price of $20 per thousand rough, $30 if dressed. Some of the mills had planers and shipped dressed lumber either green or partially air dried. Nor did softwood comprise the only shipments, as in 1878 a cargo of laurel was loaded at Westport with San Francisco as its destination.

Some early coastal shipping went elsewhere than San Francisco. Sacramento took many schooner cargoes from Albion, Fish Rock and Point Arena. Redwood lumber was now being accepted in domestic and foreign markets for interior trim, porch rails, fence posts, columns, moulding, shingles, doors, flooring, siding, shiplap, barrel staves, tank and silo stock, lath and lattice, grape stakes, railroad ties, box shook. Even the tan bark was in demand.

The Mendocino coast had its own shipyard, at Little River, for building lumber schooners. In 1875 two vessels were built in Capt. Thomas H. Peterson's yard there. Several others were completed later. The shipyard was convenient to Coombs and Perkins sawmill, being on the beach below the bluff. The ship timbers and lumber had merely to be hauled about 100 yards from the mill to the edge of the bluff, then lowered 40 feet to the beach. At Navarro, Charles Fletcher built six or eight schooners to supplement the San Francisco Bay supply.

JUMBLE OF DRUMS AND CABLES IN HIGHLINE FLYER. Used between 1926 and 1934, at Greenwood and Rockport, California, as an all-purpose logging machine. It met its limitation however when used to pull trees out by their roots.

LAWSON'S HIGHLINE FLYER. All purpose logging machine designed by Port Lawson, employee of Goodyear Redwood Company, Greenwood, California. Complicated assemblage of gears, drums, wires and pipes. Could log whole gulch in one position.

LAID EVERYTHING LOW. Line ran from ridge to ridge with radius of ¾ mile and took everything in sight. It was unsuccessfully tried in falling operations as well. Choker was set around tree as high as possible and as it fell, line would lower it slowly. (Photos Union Lumber Company Collection)

While the Hammond Lumber Co. landed 53 rafts between 1902 and 1922, redwood rafting was never a success. Capt. Hugh Robertson who freighted sea-going rafts from the Columbia River to California ports, built a redwood monster at Fort Bragg in 1891 but it could not be launched from the shipways.

The small schooner Newsboy was the first ship owned by Capt. Robert Dollar, carrying lumber from his sawmill at Usal to San Francisco. On one voyage this ship entered the Fort Bragg harbor during a gale, having been unable to make port at Usal.

Capt. Dollar was aboard, carrying a several thousand dollar payroll for his mill. When the Newsboy was halfway into the harbor, she was lifted up like a cork by a mountainous wave and literally tossed over the reef into the relatively quiet waters of the bay. Capt. Dollar came ashore, his clothes soaking wet, his face pale, but his money belt was intact. After telephoning his son Stanley in San Francisco, he hired a horse and rode all night back to Usal to pay off his men. From that time on the north reef of Fort Bragg harbor was known as "Newsboy Channel."

As sail gave way to steam more lumber producers built or acquired vessels. C. R. Johnson and his associates formed the National Steamship Co., which in time owned the three Noyos, the National City and Brunswick; chartered or operated the steam schooners Coquille River, South Coast, Higgins, Berkeley, Fort Bragg, Phoenix, Sea Foam, Arctic, Greenwood and Sequoia, the latter built at Fort Bragg. The story is told of the Kruger, chartered by Union Lumber Co. for one voyage and which came into Fort Bragg to load. She was commanded by Captain Hansen, and he was as proud of his ship as he was contemptuous of Fort Bragg harbor. While the Kruger was loading an angry "souwester" began to make up, and everybody on the wharf warned against the ship going out in such a storm. But Captain Hansen scoffed at this advice. He said the storm was "just a little blow", and that his ship could stand any sort of storm. He even intimated that the men who had warned him did not know a real ship when they saw one; that they were so accustomed to the "little tubs" which came into their "dog-hole" that they didn't realize what a good ship looked like or could withstand.

DANGER AHEAD. Lawson's Highline Flyer being used in loading operation. Even when conditions were right it was a dangerous setup with heavy carriage and extreme length of wire rope. (Photos Union Lumber Company Collection)

LOADING REDWOOD ON TRAMP STEAMER by cable and carriage off Noyo River near Fort Bragg. Harbor last used in 1938. (Photo Union Lumber Company Collection)

So Captain Hansen continued taking on lumber, and that afternoon the Kruger sailed. The storm was still raging, and kept up all night and throughout the following day. About noon of that day, those on the wharf heard four sharp whistle blasts, a sign of distress and saw the bow of the steam schooner Seafoam coming from the north, with something in tow. A few minutes later he was able to see that the tow was a terrible wreck of a ship, and it proved to be the Kruger. During the preceding night, in the heavy weather she had foundered off Caspar. Her steam had gone out, Captain Hansen and his crew abandoned ship and took to the small boats, and somehow managed to get ashore. Her deck load had carried away taking the house with it, and she was also badly waterlogged. Afterwards, with the wind blowing a terrific gale from the south, the wreck had drifted some distance north, where the Seafoam found it and put a towline aboard.

The steam schooner Brunswick happened to be tied up at the Fort Bragg wharf at the time, and her master, Captain Ellefsen, phoned San Francisco and after explaining the situation, asked for instructions. He was told to give full help to the Seafoam, which had indicated her inability to handle the wreck

PACIFIC LUMBER COMPANY—SCOTIA before fire in 1895. (Photo Mrs. L. Skelton courtesy Pacific Lumber Company) **EARLY SHINGLE MILL—SCOTIA.** (Photo H. M. Smith courtesy Pacific Lumber Company)

alone. So the Brunswick set out to help, and Captain Miller, the Seafoam's master, megaphoned the Brunswick to put a line aboard and help tow. The Brunswick did so, and began pulling. She was in the lead, with the Seafoam behind her, and with the wreck of the Kruger behind the Seafoam. There was now no one aboard the Kruger.

The "souwester" continued throughout the second night with undiminished fury, and by morning the Brunswick and Seafoam, still towing the Kruger wreck, had been able to make only about thirty miles. As it was winter, it was still dark when, at 5 a.m., the Seafoam, which was behind the Brunswick, blew four whistles and after a short interval repeated them, indicating she was in distress. As soon as there was any visibility, the Brunswick turned to, to find out the trouble. Through his megaphone, Captain Miller, the master of the Seafoam, shouted that his ship was in imminent danger of being pulled in two by the wreck and he would have to let go at once.

With heavy seas running, it took the Brunswick until afternoon to get a towline made fast to the wreck, and by then the swells had worked the ships around until they were headed west. Finally, with its towline secure, the Brunswick's master gave the word to go; but in the maneuvering required to get back on their southward course again, the wreck of the Kruger turned turtle. This naturally made the Kruger still harder to handle, and the master

SCOTIA AFTER 1895 FIRE.
(Photo Mrs. L. Skelton courtesy Pacific Lumber Company)

YARDING IN REDWOODS WITH DOLBEER DONKEY IN '90s. Note use of manila line and purchase blocks. (Caterpillar photo F. Hal Higgins Collection)

TROUBLE ON THE G.O.P. Baldwin wood burner used from 1890 to 1900 in the Mt. Shasta yellow pine country between camp at Sisson and Rainbow Mill—owned by Leland, Woods and Sheldon which later became Mt. Shasta Pine Co. G.O.P. stood for 'Get Out and Push". Engine now in Collier State Park Logging Museum, presented by Harry Benton, son of Thomas Hart Benton, who worked in his father's sawmill, drove 3-wheel steam traction engines in the woods and became head of Mt. Shasta Pine Co. (Photo George Schrader Collection, Collier State Park Logging Museum)

YARDING RED-WOODS WITH VERTICAL SPOOL DONKEY. (Photo Union Lumber Company Collection)

of the Brunswick conferred with his chief engineer as to whether the Kruger was worth salvaging. The chief said he believed the Kruger's engines were still in her and she was therefore worth salvaging, so Captain Ellefsen went back to the bridge and the Brunswick started full speed ahead. But pull as she would she still went backward, and by evening it was apparent that unless she let go the tow, she would go on the rocks at Navarro and herself be a wreck. Accordingly, the Brunswick let go the Kruger and steamed back to Fort Bragg, staying outside all night, but moving into the wharf next day to take on more lumber.

It was now nightfall, and the Brunswick, in trying to get out of the harbor in the darkness and heavy weather, struck the north reef and lost her rudder. Captain Ellefsen managed to rig up a foresail, and with this done he would alternately run his propeller and stop it. Running it would force the stern of the ship to the right, and stopping it would cause the foresail to pull the bow around.

This enabled them to keep off the rocks, but as they immediately began drifting north, it was necessary to rig a jury rudder. In doing this they somehow got a line into the propeller and jammed the mechanism. With no propeller and only a makeshift rudder, there was no way to keep from drifting farther north, but she did manage to avoid going ashore, and finally,

UNION LUMBER WOOD BURNER locomotive in early 1900's. Bob cars used link and pin couplings. (Photo Union Lumber Company Collection)

POWER SAW IN BACK CUT. Weighing up to 250 pounds, early power falling saw was first used by Union Lumber Company in 1940 but was too heavy for practical use. (Photo Union Lumber Company Collection)

S. S. MARU ON BIG RIVER AT MEN-DOCINO. Operated by the Mendocino Lumber Company from early 1900's to 1937 to bring log rafts from boom 4 miles upstream to sawmill, also to pick up sinkers and haul logs off the banks left there by high water. Crew consisted of a captain and engineer licensed for bays and rivers, and three hands. Usually made daily trips waiting for tide water to reach flood stage, then guided rafts downstream on ebb tide. Equipment consisted of a donkey engine, paddle wheel, rudder, a wheel, main line and back line, lifeboat, life preservers. (Photo Union Lumber Company Collection)

through the rain those aboard her saw lights coming toward them. It proved to be the Seafoam coming to help. But try as she did, she could not get close enough because of the high seas.

There were, of course, no ships' radios at this time, but the failure of the Brunswick to reach San Francisco so alarmed the Union Lumber Co. that Capt. Hammer of the steam schooner Phoenix was ordered from San Francisco to find the mouth of the Russian River, and put a line on the Brunswick's stern to keep her from veering. From then on this strange ocean cavalcade made much better time. However, when they got just outside the heads, at the entrance to San Francisco Bay, the Brooklyn stopped and announced that with night coming on and the seas still rough, it was too dangerous to go in, so they would have to wait outside till morning. This was too much for Captain Ellefsen and Captain Hammer. They objected so strenuously, and punctuated their objections with such strong sea language that the master of the Brooklyn decided he would yield, and they went in that night after all. Later there was a court battle over the whole affair. The Brooklyn's claim for salvage was ruled out, and her subsequent claim for a towing bill of more than $6,000 was allowed only in small part.

Union Lumber Company, for more than 50 years the major producer of lumber in Mendocino county, had its origin about 25 years after the first mill was established. In 1875 the Field Bros. built a sawmill on Mill Creek. This mill burned in 1877, but was rebuilt by the Stewart brothers and Hunter, operating as Newport Sawmill company, in 1878. In 1880 they moved the mill and reconstructed it on the South Fork of Ten Mile river.

Here in 1882, C. R. Johnson arrived and bought a part interest in the firm, which became Stewart, Hunter and Johnson. In 1884 construction was started on a new sawmill at the site of the old army post known as Fort Bragg, using parts of the Stewart, Hunter, and Johnson mill. The company was known as the Fort Bragg Redwood Company, and C. R. Johnson was its leading figure.

The mill, completed in November, 1885, was destroyed by fire in April 1888, but immediately rebuilt. The Union Lumber Company was formed in 1891 by a combination of the Fort Bragg Redwood Company, and White and Plummer. The latter company had been cutting principally ties at a sawmill on the Noyo river, originally built by A. W. Macpherson in 1858.

The newly formed Union Lumber Company built a tunnel through the ridge between Pudding Creek and the Noyo river to tap the timber in the Noyo Basin. By 1904 it had become the leading lumber producer in Mendocino county. In 1905 this company acquired the Little Valley Lumber Company mill and timber and part interest in each of the Glen Blair Lumber Company and Mendocino Lumber Company sawmills and timber holdings.

In the early '60s the Pacific Lumber Company began operations, first manufacturing being at Forrestville, now Scotia, and Field's Landing with logging up the Eel River. Bulls were used on the first skidroads but rail facilities developed rapidly. In 1893 the firm began a shipbuilding program and eventually operated a large fleet of schooners and steamers. Pacific continues, as does Hammond Lumber Company at Samoa, as one of the large factors in redwood production.

OLD HAMMOND SKIDROAD. (Photo Hammond Lumber Company Collection)

1917 GMC TRUCK WITH WOODEN WHEELED TRAILER at Kesterson Mill west of Dorris, California. Rigged with rear axle of horse logging truck. Note solid rubber tires, corner bind chains and "fid hooks". (Photo Ivan Kesterson Collection, Collier State Park Logging Museum)

CHOPPERS REST after falling big redwood.
(Photo Hammond Lumber Company Collection)

SIX-YOKE OF BULLS NEAR MENDOCINO CITY—1905 in Mendocina Lumber Co. woods. Water barrel supplied water to be thrown on skid road to make the logs slide easier. A water slinger with two five-gallon buckets on each end of a pole hung over his shoulders would run in front of the teams throwing water out of each bucket with a tomato can. He would dip the buckets into the water barrels on the run and keep on the move. (Photo Union Lumber Company Collection)

AIR SAW BUCKING YELLOW PINE at McCloud River Lumber Company in 1906. Two men fixed carriage on log, two handled air cylinder and saw. Air pumped by steam traction engine, hose reaching 600 feet. (Photo Parsons Collection, Collier State Park Logging Museum)